SURGICAL ROBOTICS

Edited by **Serdar Küçük**

Surgical Robotics
http://dx.doi.org/10.5772/65524
Edited by Serdar Küçük

Contributors

Atsushi Nishikawa, Zheng Wang, Sicong Liu, Jing Peng, Michael Chen, Do Hoon Koo, Dong Sik Bae, June Young Choi, Igor Monsellato, Fabio Priora, Maria Caterina Canepa, Vittorio D'Adamo, Giuseppe Spinoglio, Luca Matteo Lenti, Alain Sánchez González, Begona Garcia Zapirain, Rodolfo José Oviedo Barrera, Konstantinos Konstantinidis, Savvas Hirides, Michael Konstantinidis, Petros Hirides

Published by InTech
Janeza Trdine 9, 51000 Rijeka, Croatia

Publishing Process Manager Nina Kalinic
Technical Editor SPi Global
Cover InTech Design team

First published January, 2018
Printed in Croatia
Legal deposit, Croatia: National and University Library in Zagreb

Additional hard copies can be obtained from orders@intechopen.com

Surgical Robotics, Edited by Serdar Küçük
p. cm.
Print ISBN 978-953-51-3719-1
Online ISBN 978-953-51-3720-7

For EU product safety concerns:
IN TECH d.o.o., Prolaz Marije Krucifikse Kozulić 3, 51000 Rijeka, Croatia,
info@intechopen.com or visit our website at intechopen.com.

Meet the editor

Serdar Küçük received his BA and MSc degrees from Marmara University (Istanbul, Turkey) in 1995 and 1998, respectively. He received his PhD degree from Kocaeli University in Turkey. He is currently working as an associate professor at the Department of Biomedical Engineering in Kocaeli University, Turkey. He has several scientific publications including international conference papers, journal papers, books, and book chapters. He serves as reviewer of several well-known robotic journals. He has also become an editor of a scientific book. His research interests are optimization, control, kinematics, and dynamics modeling of serial and parallel robotic manipulators. Lately, he has also been interested in designing electrically controlled above-knee prosthetics and hand-wrist rehabilitation robots.

Contents

Preface

Robots are increasingly becoming popular nowadays. In the beginning, they had been used especially in mass production lines in small-scale to large-scale factories. Now, they have almost entered every field of modern human life. They are now used in several fields such as agriculture,car production, space and underwater exploration, medical purposes, hazardous and dangerous environments, and other manufacturing industries. Robots are preferred in the fields mentioned above since they perform the given tasks faster and better than humans. Precision, intelligence, and long-term working cycles make them preferable in these fields.

Robots have become also very attractive machines for medical staff recently. Robotic technology has increasingly been preferred by the medical professionals since they have been used in several clinical applications such as neurosurgery, orthopedics, radiosurgery, and cardiac surgery. Medical robots are preferred since they present better results compared to traditional methods such as smaller incision, higher accuracy, lesser recovery time, and better stability.

Medical robotic technology can be divided into three progressive generations. The first-generation robots (PUMA, SCARA, and Delta) were originally industrial robots that had been modified for performing medical applications in orthopedics, neurosurgery, radiology, and radiotherapy. These robots were used in the 1980s. The following ten years had witnessed the second-generation robotic theology that has been especially developed for executing surgical operations. The second-generation robots possessed special designs that make surgeons easily operate the whole system. They have been preferred in fields like orthopedics, radiology, and especially minimally invasive surgery. After the 2000s, the new robotic manipulators had started to develop. These robotic manipulators were called as the third-generation medical robots that had been designed for performing difficult surgical and medical operations. From the first approved surgical robot AESOP to the current da Vinci Surgical System, there have been several different kinds of surgical robots produced until now. Although the history of surgical robots is very short compared to the history of surgery, thousands of them have been installed in the hospitals worldwide, and hundreds of thousands of people have been treated by these surgical robots. Nowadays, the achievements of surgical robotics amaze both medical professionals and the patients. As a robotic professional, I believe that the future of the surgical robotics will also be spurring and astonishing. It is noteworthy to follow up on the evolution of the surgical robotics in the future.

As a last word, I would like to thank all the authors who have contributed to the book chapters with their valuable novel ideas and their knowledge on the current developments about surgical robotics.

Assoc. Prof. Serdar Küçük, PhD
Kocaeli University, Technology Faculty,
Department of Biomedical Engineering,
Turkey

Section 1

Overview of Robotics Surgery

Chapter 1

The Next-Generation Surgical Robots

Zheng Wang, Sicong Liu, Jing Peng and
Michael Zhiqiang Chen

Additional information is available at the end of the chapter

http://dx.doi.org/10.5772/67515

Abstract

The chronicle of surgical robots is short but remarkable. Within 20 years since the regulatory approval of the first surgical robot, more than 3,000 units were installed worldwide, and more than half a million robotic surgical procedures were carried out in the past year alone. The exceptionally high speeds of market penetration and expansion to new surgical areas had raised technical, clinical, and ethical concerns. However, from a technological perspective, surgical robots today are far from perfect, with a list of improvements expected for the next-generation systems. On the other hand, robotic technologies are flourishing at ever-faster paces. Without the inherent conservation and safety requirements in medicine, general robotic research could be substantially more agile and explorative. As a result, various technical innovations in robotics developed in recent years could potentially be grafted into surgical applications and ignite the next major advancement in robotic surgery. In this article, the current generation of surgical robots is reviewed from a technological point of view, including three of possibly the most debated technical topics in surgical robotics: vision, haptics, and accessibility. Further to that, several emerging robotic technologies are highlighted for their potential applications in next-generation robotic surgery.

Keywords: surgical robot, review, soft robotics, origami

1. Surgical robots today

Two decades since the American Food and Drug Administration (FDA) approved the first robotic device for surgical application, the establishments and achievements for robotic-assisted surgery are remarkable [1–3]. A brief skim through the history of surgical robotics would reveal the mileage covered in this very short period comparing with the history of surgery. The first FDA-approved surgical robot, the automated endoscopic system for optimal positioning (AESOP, Computer Motion Inc.), was a teleoperated robotic endoscopic camera that followed the commands of the surgeon via either pedals or voices. The AESOP system was successfully used in laparoscopic surgical procedures in areas such as urology, gynecology, etc., [4–7]. The subsequent ZEUS robotic system (Computer Motion Inc.) complemented an AESOP camera with two teleoperated robotic manipulators that were also continuously controlled by the surgeon through motion or voice commands [1, 8]. Despite its clinical success, the ZEUS was rivaled by the da Vinci Surgical System (Intuitive Surgical Inc.) and was discontinued two years after clearing FDA due to company merger [9, 10]. The da Vinci, on the other hand, has been the class leader for robotic-assisted surgery ever since. General laparoscopic surgery was among the first group of FDA-approved procedures for the da Vinci system in 2000, followed by radical prostatectomy in 2001, and urological surgical procedures in 2005 [11]. The list of FDA-approved procedures kept expanding, until the recent one for benign hysterectomy and salpingo-oophorectomy procedures for the latest version of the da Vinci system in 2013 and 2014 [12].

Besides expanding to new surgical areas, surgical robots have also made remarkable success in market penetration. The total number of da Vinci surgical systems installed (accumulatively) by December 2014 was 3,266 (2,223 in the US), with 570,000 procedures performed in the year 2014 [12]. Both the clinical and commercial successes have stimulated global research attention in surgical robotics. For physicians, there are various aspects of robotic-surgery–related research being investigated, ranging from efficacy [13–16] to benefits for patients [16, 17], as well as risks [18–20] and ethics [20, 21]. Another major aspect of research is surgical training, where surgical robots are generally believed to shorten the learning curve for laparoscopic surgery for young surgeons [22–25], while some variations were reported on skilled open surgeons transferring to robotic procedures [26]. Surgical training was also investigated by scientists and engineers, but via a different approach. Utilizing the complete mechanical separation between the surgeon and the patient, it was possible to generate computer signals in virtual reality (VR) and present to the surgeon using exactly the same surgeon's interface console used in real surgeries. Virtual reality surgical simulations could easily be programmed to emulate cases difficult or rare in the real world with high resemblance, hence saving animal and patient models, while significantly reducing the surgical training cost [27–29]. The VR-based surgical training was reported to be efficient in training new surgeons to robotic surgery [30, 31].

2. Technical innovations for surgical robotics

While surgeons kept innovating in robotic surgery by developing new procedures and training programs for the commercially available surgical robots, scientists and engineers have strived to innovate for robotic surgery outside the operational theater. One major direction was to develop new functionalities for the existing surgical robots. Among the various research directions, the most successfully implemented functions are vision, haptics, and accessibility.

http://dx.doi.org/10.5772/67515

2.1. Innovations for vision

In robotic laparoscopic surgeries, the surgeon no longer has a direct view of the surgical site, but must rely on camera images displayed on computer screens. Before the age of high-definition video, this used to be a significant limiting factor such that the surgeon did not have a view of the surgical site with sufficient resolution. This concern was soon overcome by high-definition high-quality live video streaming, even three-dimensional (3D), which are already standard specifications for many available surgical robots [12]. The benefit of using cameras did not end with stereo vision. Making use of advanced lens systems, the surgeon could have an artificial view of the surgical site beyond the capability of the naked eye, for instance, the ultra-wide angle fisheye view from an endoscope or a super macro enlarged view of a tiny area otherwise not visible to a human. Moreover, since the video presented to the surgeon was in fact a computerized image sequence, it was possible to overlay a variety of information and other images [32, 33]. The resulting augmented vision has already been successfully implemented in surgical robots for the surgeon's maximum benefits [34]. Furthermore, overlaying preoperative imaging results and even live imaging data such as ultrasound or magnetic resonance imaging (MRI) could potentially solve the navigation challenge for laparoscopic surgery. Pioneering systems have already been reported for both preoperative and intraoperative imaging augmentation [33, 35–37].

2.2. Innovations for haptics

Another major and yet still ongoing debate is on whether haptics is a necessity for robotic surgery [34]. The term haptics has been used to refer to the sense of touch in general, while in this context, it only refers to providing force feedback signal to the surgeon on the surgeon's console, so that the surgeon could feel how much force is being applied even without direct view over the contact point, for better and safer handling of tissues [38]. Haptics of the same narrow sense had been investigated for a much longer period of time in general robotics research. Controlling forces at the interaction point had been studied in the 1970s [39, 40], with hybrid force/position control algorithms proposed in the late 1970s and the early 1980s [41, 42]. Soon afterwards, the concept of impedance control was formulated in the mid-1980s, where the virtual stiffness of a robotic manipulator could be controlled instead of position or force individually, to cope with any unpredicted interaction status [43, 44]. This concept quickly became one of the most popular and well-established control approaches in robotics until today [45]. By the time of the first-ever FDA approval on surgical robots (the AESOP), roboticists proposed the concept of transparency: that an ideal teleoperation system should be transparent to the user, such that every command could be faithfully executed and every event in the remote environment could be fed back to the user [46, 47]. All of the above concepts were built on available and high-quality real-time force feedback signals, which roboticists took for granted. Unfortunately for surgical robots, it was not the case. Due to strict spatial constraints, there was no force sensor available at that time that could fit into the instruments, hence the first generation of surgical robots was not equipped with force sensors, and naturally there was no force feedback [34, 38].

While engineers could not get over the fact that the state-of-the-art surgical robots were still utilizing the pre-1980 technology without proper force sensing, surgeons were starting to be trained to use the haptic-less surgical robots and estimate interacting forces by visual information [48, 49]. After the remarkable clinical achievements of haptic-less surgical robots, the addition of haptics to existing surgical robots became a radical move, in the eyes of the very group of surgeons who were radical enough to adopt robotic surgery earlier. In fact, this makes the underlying argument for the majority of literature against haptics in robotic surgery: since the current robots are already so good without it, if the additional complexity, unknown risks, and added costs could still be justified [49–51]. This hesitation was caused, at least partially, by technical reasons: in early surgical robotic systems, haptic feedback was either patched on or estimated/simulated, the performance of which was rather limited, hence surgeons were less in favor of the outcomes [48]. However, with the fast developments in robotic technology, recent surgical systems with haptic feedback are equipped with new force sensors and very well implemented control [52, 53], and as a result, more and more studies showed that haptic feedback became one of the most wanted features for the next generation of surgical robots [54–59].

2.3. Innovations for accessibility: SIL and NOTES

Another important area of technical innovation is accessibility. One of the main improvements laparoscopy had over open surgery was the significantly reduced size of incisions; hence, the alias "minimally invasive surgery" became more familiar to the general public. Reducing the incisions resulted not only in cosmetic improvements but also in a spectrum of procedural and postoperative benefits to both the surgeon and the patient [60–62]. However, surgeons had to undergo specific training with a steep learning curve to accommodate the compromised vision and maneuverability [63, 64]. This was precisely what the first generation of surgical robots took on manual laparoscopy, removing the burden of maneuverability from the surgeon by automatic control programs and electric motors, such that the surgeon no longer needed to think about the small incisions or apply fatiguing excessive forces, but focus on the surgical procedure [23, 65]. As a result, the learning curve for robotic laparoscopic surgery is much shorter [22–25]. While manual laparoscopy is still a required training, there have been studies in comparing the use of surgical robots by surgeons experience or inexperience with manual laparoscopy [26, 66, 67].

With the clinical and general adoption for laparoscopic surgical robots, roboticists tackled the more challenging single-incision surgery (SIL), where the multiple small incisions in laparoscopic surgery were further merged into one. The idea of SIL was first proposed as a manual procedure, and grew into a daily surgical routine for general surgery in particular, especially for transoral, transanal, and transvaginal interventions [68–70]. The majority of manual SIL procedures were carried out using a single instrument for intervention, as laparoscopic SIL was found with compromised practicality, where the surgeon had to either reverse the motion of the instrument tips or cross his/her own hands to accommodate the immobilizing incision point, being a very counterintuitive exhaustive motoring task to add to the mental burden for the surgeon [71]. However, various studies have pointed out that, after proper training, the efficacy for laparoscopic SIL is at least as good as standard laparoscopy [72–74]. Robotic technology bares every potential to overcome the primary limiting factor for SIL: constant and high mental burden of motoring control for the surgeon. Assuming sufficient instrumental rigidity and maneuverability, the automatic control program could drive the robotic instruments around one incision in the same way as driving them around multiple incisions. This, however, requires redesigning the hardware to provide the necessary kinematic structures for the additional complexity in motion mapping. Single site surgical robotic system has already been released, and will be accumulating clinical results in the near future [12, 75–78].

In parallel with laparoscopic SIL, another approach to increasing accessibility is by introducing robotic technology to flexible endoscopy. Endoscopic interventions are slowly growing popular after the introduction of endoscopic submucosal dissection (ESD) by Japanese physicians [79]. ESD was first targeted at endoscopic removal of neoplasia or early-stage gastric cancer [80, 81]. The technique could potentially unify the imaging, diagnostic, and treatment procedures, and find the basis for natural orifice transluminal endoscopic surgery (NOTES) [82–84]. However, in practice, manual ESD required extensive training and experience, and remained technically challenging to execute for both surgeons and endoscopists [85–87]. Overcoming the technical hurdle, the first endoscopic surgical robot was introduced by enabling multiple degrees-of-freedom (DOF) triangulated instrumentation on a standard endoscope platform [88]. The robot adopted the master-slave design similar to laparoscopic surgical robots [89, 90], and was enabled with haptic feedback [91, 92]. Robotic ESD was the first targeted procedure, with a series of porcine model [93, 94] and human trials [95], followed by a preclinical trial on full-thickness mucosa removal [96]. Behind the clinical success, significant engineering efforts were spent overcoming the cable transmission issues under very tight spatial constraints for the endoscopic instrument channels, where mechanical transmission [97, 98], static [99, 100] and dynamic [101, 102] friction attenuations were investigated thoroughly to improve the performance of the robot under the harsh working environments of the endoscope for both ESD and NOTES [103].

2.4. Global attention and trends in surgical robots

The success of laparoscopic and endoscopic surgical robots had stimulated worldwide attentions in surgical robot research, for instance, the laparoscopic telesurgical RAVEN robot [104, 105], the Magellan endovascular robot [106], snake-like surgical robots [107, 108], MRI-compatible surgical robots [109, 110], single-incision laparoscopic robots [111–116], and endoscopic robots [117–121].

The first observation is the global flourish of surgical robot research. The non-exhaustive country list includes the US, the UK, Germany, Italy, China, Japan, Korea, and Singapore. The cited works here did not include literature published in non-English format, or industrial developments, which could be expected considering the strong application orientation for this field. The second trend is the clear convergence of targeted surgical procedures for the various, independently developed surgical robotic systems. While earlier systems such as the RAVEN [104, 105] was still designed for laparoscopic surgery with multiple incisions, later laparoscopic robots were all aimed for single-incision procedures [107–116]. For endoscopic alternatives, nearly all systems were aimed fully or partially at NOTES [118–122]. General surgery and urologic surgery were the most common two surgical areas mentioned in the system development goals. The third observation is the technology used in the new systems. All of the cited systems used cable transmission to remotely drive the robotic end-effector except one design that utilized a screw-drive [117]. To create the cable-pulling motion, various techniques were employed; the majority used electric motor [103, 106–109, 113–115, 119–121], while others used shape-memory alloy [116, 118], pneumatics [109], piezoelectric actuator [110], and magnetics [117]. The final observation is on the manipulator structure. Both SIL- and NOTES-oriented surgical robots are attempted to integrate multiple (three to six) DOF mechanisms under a very tight spatial constraint, while required to deliver high gripping force for tissue handling and suturing tasks. While conventional revolute joints were still employed in some designs [122], articulated and continuum mechanisms were the clear trend for their better integration potential, stronger structure, and higher force capabilities [123]. The kinematic designs of typical surgical robots were reviewed in Ref. [124].

3. Emerging technologies for future surgical robotic applications

Robotic research in general is also moving at remarkable speeds. There are constantly new developments and discoveries that could potentially be translated into surgical robotic applications. Here, two of the emerging new technologies are highlighted: origami and soft robotics. Both directions are quickly picking up momentum in recent years, with the potential to tackle on one of the fundamental challenges in surgical robotics, and both already had pioneering systems being reported for related applications.

3.1. Origami in surgical robots

Origami is the art of intricately folding a sheet of paper into elaborate 3D sculptures and objects [125]. The essential elements of an origami pattern are the facets and crease lines (mountain and valley folds) that formed flat facets, i.e., quadrilaterals or triangles, and fold lines which are considered as revolute hinges connecting the facets. As a result, origami mechanisms could be folded from 2D states to 3D structures, such as the Miura-ori patterned sheet [126] and deployable structures [127, 128]. By implementing actuation in the hinges, self-folding origami composed of shape-memory polymer [129] and print-and-self-fold miniature electric devices could be obtained [130].

Origami mechanisms have the potential to tackle two crucial challenges faced by surgical robots: fabrication and assembly. A micro-fabrication technique known as Pop-Up Book MEMS [131] could create 3D, multi-material, and monolithic meso- and microstructures using purely 2D planar manufacturing and origami folding techniques [132]. The Pop-Up technology allows for the fabrication of complex, multifunctional electromechanical devices on the 0.1–10 mm scale, significantly below the size limitation for traditional machining techniques. It consists of flexible (polyimide), structural (carbon fiber or metal), and adhesive layers. To overcome planar limitations inherent to MEMS, surface-machined pin-and-staple hinges [132] and polymer flexures [133] are used to create folding linkages.

In addition to the fabrication scale advantages, origami mechanisms also allow for novel assembly possibilities. As the boundary of miniature surgical instruments keeps being pushed, the difficulty for the assembly, bonding, and packaging processes would increase in multifolds. Self-folding (self-assembly/self-deployable) origami-inspired miniature devices have been demonstrated to effectively solve the assembly challenge [132]. A series of self-folding grippers have been demonstrated in Refs. [134–136] with a variety of materials, shapes, and sizes, mostly targeted at single-cell manipulation. Techniques such as photolithography, electron-beam lithography, and soft lithography have been used to precisely pattern two-dimensional sheets of materials, namely metals, semiconductors, and polymeric films. Actuations derived from surface tension, residual stress, thermal or PH stimuli are used to fold patterned sheets into three-dimensional structures [137]. Instruments of an SIL surgical robot have a much larger scale than the cell manipulators above, while also requiring much higher forces. A Pop-Up-based surgical robot grasper was developed as given in Ref. [138]. Besides easy assembly, a novel feature was the integrated force sensing during the same fabrication and assembling procedure.

Besides the Pop-Up-based grasper in Ref. [138], another grasper design based on origami mechanism was reported in Ref. [139] with four DOF and was actuated by shape memory alloy (SMA). Origami could eventually revolutionize surgical instrument design and manufacturing, with self-assembling micro-scale robotic end-effectors integrated with sensors and actuators. Moreover, the actuator could be delivered into the surgical site in 2D form and self-assemble into 3D working form afterwards.

3.2. Soft robotics for surgical applications

Soft robotics is another rapidly emerging research field. Soft robots are commonly fabricated with flexible and elastomeric materials to achieve complex motions with simple mechanical structures [140, 141]. Generating motions without relying on rigid structures or components, these systems are ideal for bio-mimicking [142, 143] and manipulating delicate objects [144, 145]. Soft robots could be actuated with electrical charges [146], chemical reactions [147], and most commonly pressurized fluids [143, 144, 148, 149]. When pneumatic/hydraulic soft robots are pressurized, the internal fluid chambers would expand and deform the actuator. By selectively controlling and redirecting the deformation, multiple forms of motions could be created or even combined, such as contraction/extension [150], bending [143, 144, 148, 151–153], and twisting [142, 154]. Soft robots have a long list of desirable features, such as low weight, high power-to-weight ratio, low material cost, and ease of fabrication [141, 142].

For surgical robotic applications, soft robots have one clear advantage: inherent compliance. Without any rigid component, the entire robot is soft and compliant at rest. Even after pressurization, its soft structure and fluidic actuation media would still allow some level of compliance and back-drivability under extreme conditions [142]. This inherent compliance translates to safe and atraumatic tissue handling and manipulation during surgical procedures. With the vast majority of the current instruments for surgical robots made from metal or other high-stiffness materials, soft robots bear the potential to offer soft alternatives for specific situations. A soft robotic grasper was developed for atraumatic tissue handling in robotic surgery, as a safe interface between the rigid surgical instrument and the delicate human organ [155]. The preliminary results were very promising for the future application of soft robotics into surgical systems.

4. Conclusions

Technology had once again brought a paradigm shift into operational theaters toward robotic surgery. Robotic surgery has been and will continue to be one of the fastest growing fields in medicine in the foreseeable future. On the other hand, as elaborated in this article, the current generation of surgical robots is far from perfect in the sense of robotic technology, neither are they providing the surgeons with the ideal user experience. This is in part due to the inherent conservation in medical innovation, such that only the well-matured and proven technologies could penetrate the regulatory barrier into implementation. Another important reason not to be overlooked is the exploration and make-do spirits of visionary surgeons: it is not unusual that surgical robots are experimented in new procedures or even surgical areas it was not originally designed for. Regulatory would also put efficacy and safety over the surgeon's user experience as the main considerations, as they are directly related to the benefits of the patient, the regarded real end user for surgical robots. Therefore, as long as the (previously approved) surgical robot could be used in a new procedure effectively and safely, it could potentially be approved for clinical practice.

Built on the remarkable success of current surgical robots, in the near future, there will be a spectrum of new surgical robots, developed by both robotic laboratories and companies all around the globe, and employing a wide range of novel technologies, including the ones introduced in this article. The majority of such new systems will strive to reduce both the footprint of the robot and the size of the incision, for better suitability for SIL and/or NOTES. Automated surgery would still be a challenging area as, until now, the judgments of the surgeon remained the core of the entire surgical procedure. Shifting the role of robots from assistive instruments and operational interfaces to decision makers, even partially, would require a much greater effort, both in research/development and in the mentality of surgery, than technically improving surgical robots within their current range of responsibility. However, both the acquisition cost of the robotic system and the maintenance and procedural costs will be lowered, even if this means compromising the generalizability and introducing new robots more specialized in certain surgical areas or procedures. This would help in promoting robotic surgery into regional and specialized clinics. On the other hand, given the complexity of the design iteration and the time required for the regulatory approval procedure, the development of new surgical robot systems would hardly catch up with the speed of pushing new surgical boundaries. For this, surgeons and roboticists will continue to innovate based on the current generation of surgical robots, add new functions, develop evolutionary updates, apply modifications to fit new procedures, as well as compose new training protocols and programs to fully cultivate the potentials of surgeons.

Author details

Zheng Wang[1,2]*, Sicong Liu[3], Jing Peng[1] and Michael Zhiqiang Chen[1,2]

*Address all correspondence to: zwangski@hku.hk

1 Department of Mechanical Engineering, The University of Hong Kong, Hong Kong

2 HKU Shenzhen Institute of Research and Innovation (SIRI), Shenzhen, China

3 School of Mechanical and Aerospace Engineering, Nanyang Technological University, Singapore

References

[1] Stoianovici D. Robotic surgery. World Journal of Urology. 2000;**18**(4):289-295

[2] Camarillo D B, Krummel T M, Salisbury J K. Robotic technology in surgery: Past, present, and future. The American Journal of Surgery. 2004;**188**(4):2-15

[3] Gomes P. Surgical robotics: Reviewing the past, analysing the present, imagining the future. Robotics and Computer-Integrated Manufacturing. 2011;**27**(2):261-266

[4] Kavoussi L R, Moore R G, Adams J B, et al. Comparison of robotic versus human laparoscopic camera control. The Journal of Urology. 1995;**154**(6):2134-2136

[5] Partin A W, Adams J B, Moore R G, et al. Complete robot-assisted laparoscopic urologic surgery: A preliminary report. Journal of the American College of Surgeons. 1995;**181**(6):552-557

[6] Mettler L, Ibrahim M, Jonat W. One year of experience working with the aid of a robotic assistant (the voice-controlled optic holder AESOP) in gynaecological endoscopic surgery. Human Reproduction. 1998;**13**(10):2748-2750

[7] Alessandrini M, De Padova A, Napolitano B, et al. The AESOP robot system for video-assisted rigid endoscopic laryngosurgery. European Archives of Oto-Rhino-Laryngology. 2008;**265**(9):1121-1123

[8] Sim H G, Yip S K H, Cheng C W S. Equipment and technology in surgical robotics. World Journal of Urology. 2006;**24**(2):128-135

[9] Sung G T, Gill I S. Robotic laparoscopic surgery: A comparison of the da Vinci and Zeus systems. Urology. 2001;**58**(6):893-898

[10] Wedmid A, Llukani E, Lee D I. Future perspectives in robotic surgery. BJU International. 2011;**108**(6b):1028-1036

[11] Intuitive Surgical Inc. Annual Report 2010, Intuitive Surgical, www.intuitivesurgical.com

[12] Intuitive Surgical Inc. Annual Report 2014, Intuitive Surgical, www.intuitivesurgical.com

[13] Wexner S D, Bergamaschi R, Lacy A, et al. The current status of robotic pelvic surgery: Results of a multinational interdisciplinary consensus conference. Surgical Endoscopy. 2009;**23**(2):438-443

[14] Joyce D, Morris-Stiff G, Falk G A, et al. Robotic surgery of the pancreas. World Journal of Gastroenterology: WJG. 2014;**20**(40):14726

[15] Pai A, Melich G, Marecik S J, et al. Current status of robotic surgery for rectal cancer: A bird's eye view. Journal of Minimal Access Surgery. 2015;**11**(1):29

[16] Araujo S E A, Seid V E, Klajner S. Robotic surgery for rectal cancer: Current immediate clinical and oncological outcomes. World Journal of Gastroenterology: WJG. 2014;**20**(39):14359

[17] Kaye D R, Mullins J K, Carter H B, et al. Robotic surgery in urological oncology: Patient care or market share?. Nature Reviews Urology. 2015;**12**(1):55-60

[18] Zorn K C, Gautam G, Shalhav A L, et al. Training, credentialing, proctoring and medicolegal risks of robotic urological surgery: Recommendations of the society of urologic robotic surgeons. The Journal of Urology. 2009;**182**(3):1126-1132

[19] Weinstein G S, O'Malley Jr B W, Desai S C, et al. Transoral robotic surgery: does the ends justify the means?. Current Opinion in Otolaryngology & Head and Neck Surgery. 2009;**17**(2):126-131

[20] Sarlos D, Kots L V, Stevanovic N, et al. Robotic hysterectomy versus conventional laparoscopic hysterectomy: Outcome and cost analyses of a matched case–control study. European Journal of Obstetrics & Gynecology and Reproductive Biology. 2010;**150**(1):92-96

[21] Larson J A, Johnson M H, Bhayani S B. Application of surgical safety standards to robotic surgery: Five principles of ethics for nonmaleficence. Journal of the American College of Surgeons. 2014;**218**(2):290-293

[22] Schreuder H W R, Wolswijk R, Zweemer R P, et al. Training and learning robotic surgery, time for a more structured approach: A systematic review. BJOG: An International Journal of Obstetrics & Gynaecology. 2012;**119**(2):137-149

[23] Brinkman W M, Schout B M A, Rietbergen J B, et al. Training robotic surgery in urology: experience and opinions of robot urologists. The International Journal of Medical Robotics and Computer Assisted Surgery. 2015;**11**(3):308-318

[24] Buffi N, Van Der Poel H, Guazzoni G, et al. Methods and priorities of robotic surgery training program. European Urology. 2014;**65**(1):1

[25] Honaker M D, Paton B L, Stefanidis D, et al. Can robotic surgery be done efficiently while training residents?. Journal of surgical education. 2015;**72**(3):377-380

[26] Doumerc N, Yuen C, Savdie R, et al. Should experienced open prostatic surgeons convert to robotic surgery? The real learning curve for one surgeon over 3 years. BJU International. 2010;**106**(3):378-384

[27] Van der Meijden O A J, Schijven M P. The value of haptic feedback in conventional and robot-assisted minimal invasive surgery and virtual reality training: a current review. Surgical Endoscopy. 2009;**23**(6):1180-1190

[28] Gallagher A G, Ritter E M, Champion H, et al. Virtual reality simulation for the operating room: proficiency-based training as a paradigm shift in surgical skills training. Annals of Surgery. 2005;**241**(2):364

[29] Aggarwal R, Ward J, Balasundaram I, et al. Proving the effectiveness of virtual reality simulation for training in laparoscopic surgery. Annals of Surgery. 2007;**246**(5):771-779

[30] Lerner M A, Ayalew M, Peine W J, et al. Does Training on a Virtual Reality Robotic Simulator Improve Performance on the da Vinci® Surgical System? Journal of Endourology. 2010;**24**(3):467-472

[31] Kenney P A, Wszolek M F, Gould J J, et al. Face, content, and construct validity of dV-trainer, a novel virtual reality simulator for robotic surgery. Urology. 2009;**73**(6):1288-1292

[32] Volonté F, Pugin F, Bucher P, et al. Augmented reality and image overlay navigation with OsiriX in laparoscopic and robotic surgery: not only a matter of fashion. Journal of Hepato-biliary-pancreatic Sciences. 2011;**18**(4):506-509

[33] Ukimura O, Gill I S. Image-fusion, augmented reality, and predictive surgical navigation. Urologic Clinics of North America. 2009;**36**(2):115-123

[34] Tan G Y, Goel R K, Kaouk J H, et al. Technological advances in robotic-assisted laparoscopic surgery. Urologic Clinics of North America. 2009;**36**(2):237-249

[35] Gill I S, Ukimura O. Thermal energy-free laparoscopic nerve-sparing radical prostatectomy: one-year potency outcomes. Urology. 2007;**70**(2):309-314

[36] Bos J, Steinbuch M, Kunst HPM. Design of a new image-guided surgical robot for precision bone drilling in the lateral skull base. Journal of Neurological Surgery B. 2016;**77**(S 02): FP-20-04

[37] Bowthorpe M, Tavakoli M. Generalized predictive control of a surgical robot for beating-heart surgery under delayed and slowly-sampled ultrasound image data. IEEE Robotics and Automation Letters, 2016;**1**(2):892-899

[38] Okamura A M. Haptic feedback in robot-assisted minimally invasive surgery. Current Opinion in Urology. 2009;**19**(1):102

[39] Whitney D E. Force feedback control of manipulator fine motions. Journal of Dynamic Systems, Measurement, and Control. 1977;**99**(2):91-97

[40] Patarinski S P, Botev R G. Robot force control: a review. Mechatronics. 1993;**3**(4):377-398

[41] Craig J J, Raibert M H. A systematic method of hybrid position/force control of a manipulator. Computer Software and Applications Conference, 1979. Proceedings. COMPSAC 79. The IEEE Computer Society's Third International. 1979:446-451

[42] Mason M T. Compliance and force control for computer controlled manipulators. IEEE Transactions on Systems, Man and Cybernetics. 1981;**11**(6):418-432

[43] Hogan N. Impedance control: An approach to manipulation: Part II—Implementation. Journal of Dynamic Systems, Measurement, and Control. 1985;**107**(1):8-16

[44] Anderson R, Spong M W. Hybrid impedance control of robotic manipulators. IEEE Journal of Robotics and Automation. 1988;**4**(5):549-556

[45] Buchli J, Stulp F, Theodorou E, et al. Learning variable impedance control. The International Journal of Robotics Research. 2011;**30**(7):820-833

[46] Lawrence D A. Stability and transparency in bilateral teleoperation. IEEE Transactions on Robotics and Automation. 1993;**9**(5):624-637

[47] Yokokohji Y, Yoshikawa T. Bilateral control of master-slave manipulators for ideal kinesthetic coupling-formulation and experiment. IEEE Transactions on Robotics and Automation. 1994;**10**(5):605-620

[48] Okamura A M. Methods for haptic feedback in teleoperated robot-assisted surgery. Industrial Robot: An International Journal. 2004;**31**(6):499-508

[49] Van der Meijden O A J, Schijven M P. The value of haptic feedback in conventional and robot-assisted minimal invasive surgery and virtual reality training: a current review. Surgical Endoscopy. 2009;**23**(6):1180-1190

[50] Panchulidze I, Berner S, Mantovani G, et al. Is haptic feedback necessary to microsurgical suturing? Comparative study of 9/0 and 10/0 knot tying operated by 24 surgeons. Hand Surgery. 2011;**16**(01):1-3

[51] Lanfranco A R, Castellanos A E, Desai J P, et al. Robotic surgery: a current perspective. Annals of Surgery. 2004;**239**(1):14

[52] Yamamoto T, Abolhassani N, Jung S, et al. Augmented reality and haptic interfaces for robot-assisted surgery. The International Journal of Medical Robotics and Computer Assisted Surgery. 2012;**8**(1):45-56

[53] Wang Z, Sun Z, Phee S J. Haptic feedback and control of a flexible surgical endoscopic robot. Computer Methods and Programs in Biomedicine. 2013;**112**(2):260-271

[54] Zhou M, Tse S, Derevianko A, et al. Effect of haptic feedback in laparoscopic surgery skill acquisition. Surgical Endoscopy. 2012;**26**(4):1128-1134

[55] Koehn J K, Kuchenbecker K J. Surgeons and non-surgeons prefer haptic feedback of instrument vibrations during robotic surgery. Surgical endoscopy. 2015;**29**(10):2970-2983

[56] Kranzfelder M, Schneider A, Fiolka A, et al. What Do We Really Need? Visions of an Ideal Human–Machine Interface for NOTES Mechatronic Support Systems From the View of Surgeons, Gastroenterologists, and Medical Engineers. Surgical Innovation. 2015;**22**(4):432-440

[57] Kume K, Sakai N, Goto T. Development of a novel endoscopic manipulation system: the Endoscopic Operation Robot ver. 3. Endoscopy. 2015;**47**(09):815-819

[58] Kim U, Seok DY, Kim YB, Lee DH, Choi HR. Development of a grasping force-feedback user interface for surgical robot system. 2016 IEEE/RSJ International Conference on Intelligent Robots and Systems. 2016;845-850

[59] Munawar A, Fischer G. A Surgical robot teleoperation framework for providing haptic feedback incorporating virtual environment-based guidance. Front Robot AI. 2016;**3**:47

[60] Zhu J H, Li W, Yu K, et al. New strategy during complicated open appendectomy: Convert open operation to laparoscopy. World Journal of Gastroenterology: WJG. 2014;**20**(31):10938

[61] Gaillard M, Tranchart H, Dagher I. Laparoscopic liver resections for hepatocellular carcinoma: current role and limitations. World Journal of Gastroenterology: WJG. 2014;**20**(17):4892

[62] Limongelli P, Vitiello C, Belli A, et al. Costs of laparoscopic and open liver and pancreatic resection: A systematic review. World Journal of Gastroenterology: WJG. 2014;**20**(46):17595

[63] Cadeddu J A, Wolfe J S, Nakada S, et al. Complications of laparoscopic procedures after concentrated training in urological laparoscopy. The Journal of Urology. 2001;**166**(6):2109-2111

[64] Kaiser A M. Evolution and future of laparoscopic colorectal surgery. World Journal of Gastroenterology: WJG. 2014;**20**(41):15119

[65] Sarle R, Tewari A, Shrivastava A, et al. Surgical robotics and laparoscopic training drills. Journal of Endourology. 2004;**18**(1):63-67

[66] Ugarte D A, Etzioni D A, Gracia C, et al. Robotic surgery and resident training. Surgical Endoscopy. 2003;**17**(6):960-963

[67] Schreuder H W R, Persson J E U, Wolswijk R G H, et al. Validation of a novel virtual reality simulator for robotic surgery. The Scientific World Journal. 2014;**2014**:1-10

[68] Maggiori L, Gaujoux S, Tribillon E, et al. Single-incision laparoscopy for colorectal resection: a systematic review and meta-analysis of more than a thousand procedures. Colorectal Disease. 2012;**14**(10):e643-e654

[69] Mittermair C, Schirnhofer J, Brunner E, et al. Single port laparoscopy in gastroenterology and hepatology: A fine step forward. World Journal of Gastroenterology: WJG. 2014;**20**(42):15599

[70] Cianchi F, Staderini F, Badii B. Single-incision laparoscopic colorectal surgery for cancer: State of art. World Journal of Gastroenterology: WJG. 2014;**20**(20):6073

[71] Santos B F, Reif T J, Soper N J, et al. Effect of training and instrument type on performance in single-incision laparoscopy: results of a randomized comparison using a surgical simulator. Surgical Endoscopy. 2011;**25**(12):3798-3804

[72] Farach S M, Danielson P D, Chandler N M. Impact of experience on quality outcomes in single-incision laparoscopy for simple and complex appendicitis in children. Journal of pediatric surgery. 2015;**50**(8):1364-1367

[73] Weiss H G, Brunner W, Biebl M O, et al. Wound complications in 1145 consecutive transumbilical single-incision laparoscopic procedures. Annals of Surgery. 2014;**259**(1):89-95

[74] Yun J A, Yun S H, Park Y A, et al. Single-incision laparoscopic right colectomy compared with conventional laparoscopy for malignancy: assessment of perioperative and short-term oncologic outcomes. Surgical Endoscopy. 2013;**27**(6):2122-2130

[75] Kaouk J H, Goel R K, Haber G P, et al. Robotic single-port transumbilical surgery in humans: initial report. BJU International. 2009;**103**(3):366-369

[76] Canes D, Desai M M, Aron M, et al. Transumbilical single-port surgery: evolution and current status. European Urology. 2008;**54**(5):1020-1030

[77] Kaouk JH, Haber GP, Autorino R, et al. A novel robotic system for single-port urologic surgery: First clinical investigation. European Urology. 2014;**66**(6):1033-1043

[78] Holsinger FC. A flexible, single-arm robotic surgical system for transoral resection of the tonsil and lateral pharyngeal wall: Next-generation robotic head and neck surgery. Laryngoscope. 2016;**126**(4):864-869

[79] Yahagi N, Fujishiro M, Kakushima N, et al. Endoscopic submucosal dissection for early gastric cancer using the tip of an electrosurgical snare (thin type). Digestive Endoscopy. 2004;**16**(1):34-38

[80] Gotoda T, Yamamoto H, Soetikno R M. Endoscopic submucosal dissection of early gastric cancer. Journal of Gastroenterology. 2006;**41**(10):929-942

[81] Tanaka S, Oka S, Kaneko I, et al. Endoscopic submucosal dissection for colorectal neoplasia: possibility of standardization. Gastrointestinal Endoscopy. 2007;**66**(1):100-107

[82] Sebastian G, DeMaria E J, Reynolds J D, et al. New developments in surgery: natural orifice transluminal endoscopic surgery (NOTES). Archives of Surgery. 2007;**142**(3):295-297

[83] McGee M F, Rosen M J, Marks J, et al. A primer on natural orifice transluminal endoscopic surgery: building a new paradigm. Surgical Innovation. 2006;**13**(2):86-93

[84] Rao G V, Reddy D N, Banerjee R. NOTES: human experience. Gastrointestinal Endoscopy Clinics of North America. 2008;**18**(2):361-370

[85] Deprez P H, Bergman J J, Meisner S, et al. Current practice with endoscopic submucosal dissection in Europe: position statement from a panel of experts. Endoscopy. 2010;**42**(10):853-858

[86] Teoh A Y B, Chiu P W Y, Wong S K H, et al. Difficulties and outcomes in starting endoscopic submucosal dissection. Surgical Endoscopy. 2010;**24**(5):1049-1054

[87] Berr F, Ponchon T, Neureiter D, et al. Experimental endoscopic submucosal dissection training in a porcine model: learning experience of skilled Western endoscopists. Digestive Endoscopy. 2011;**23**(4):281-289

[88] Phee S J, Low S C, Sun Z L, et al. Robotic system for no-scar gastrointestinal surgery. The International Journal of Medical Robotics and Computer Assisted Surgery. 2008;**4**(1):15-22

[89] Phee S J, Low S C, Huynh V A, et al. Master and slave transluminal endoscopic robot (MASTER) for natural orifice transluminal endoscopic surgery (NOTES). Engineering in Medicine and Biology Society, 2009. EMBC 2009. Annual International Conference of the IEEE. IEEE. 2009:1192-1195

[90] Phee S J, Sun Z, Wang Z, et al. The future of transluminal surgery. Expert review of medical devices. 2011;**8**(6):669-671

[91] Sun Z, Wang Z, Phee S J. Towards haptics enabled surgical robotic system for NOTES. 2011 IEEE Conference on Robotics, Automation and Mechatronics (RAM), IEEE, 2011:229-233

[92] Wang Z, Sun Z, Phee S J. Haptic feedback and control of a flexible surgical endoscopic robot. Computer Methods and Programs in Biomedicine. 2013;**112**(2):260-271

[93] Ho K Y, Phee S J, Shabbir A, et al. Endoscopic submucosal dissection of gastric lesions by using a Master and Slave Transluminal Endoscopic Robot (MASTER). Gastrointestinal Endoscopy. 2010;**72**(3):593-599

[94] Wang Z, Phee S J, Lomanto D, et al. Endoscopic submucosal dissection of gastric lesions by using a master and slave transluminal endoscopic robot: an animal survival study. Endoscopy. 2012;**44**(7):690-694

[95] Phee S J, Reddy N, Chiu P W Y, et al. Robot-assisted endoscopic submucosal dissection is effective in treating patients with early-stage gastric neoplasia. Clinical Gastroenterology and Hepatology. 2012;**10**(10):1117-1121

[96] Chiu P W Y, Phee S J, Wang Z, et al. Feasibility of full-thickness gastric resection using master and slave transluminal endoscopic robot and closure by overstitch: a preclinical study. Surgical Endoscopy. 2014;**28**(1):319-324

[97] Wang Z, Sun Z, Phee S J. Modeling tendon-sheath mechanism with flexible configurations for robot control. Robotica. 2013;**31**(07):1131-1142

[98] Sun Z, Wang Z, Phee S J. Elongation modeling and compensation for the flexible tendon--sheath system. IEEE/ASME Transactions on Mechatronics. 2014;**19**(4):1243-1250

[99] Sun Z, Wang Z, Phee S J. Modeling and motion compensation of a bidirectional tendon-sheath actuated system for robotic endoscopic surgery. Computer Methods and Programs in Biomedicine. 2015;**119**(2):77-87

[100] Sun Z, Wang Z, Phee S J. Haptic modeling of stomach for real-time property and force estimation. Journal of Mechanics in Medicine and Biology. 2013;**13**(03):1-25

[101] Do T N, Tjahjowidodo T, Lau M W S, et al. Dynamic friction-based force feedback for tendon-sheath mechanism in notes system. International Journal of Computer and Electrical Engineering. 2014;**6**(3):252-258

[102] Do T N, Tjahjowidodo T, Lau M W S, et al. An investigation of friction-based tendon sheath model appropriate for control purposes. Mechanical Systems and Signal Processing. 2014;**42**(1):97-114

[103] Wang Z, Phee S J, Wong J, et al. Development of a robotic platform for natural orifice transluminal endoscopic surgery. Gastrointestinal Intervention. 2012;**1**(1):40-42

[104] Lum M J H, Friedman D C W, Sankaranarayanan G, et al. The RAVEN: Design and validation of a telesurgery system. The International Journal of Robotics Research. 2009;**28**(9):1183-1197

[105] Rosen J, Lum M, Sinanan M, et al. Raven: Developing a surgical robot from a concept to a transatlantic teleoperation experiment. Surgical Robotics. Springer US, 2011:159-197

[106] Hemmerling T M, Taddei R, Wehbe M, et al. First robotic ultrasound-guided nerve blocks in humans using the Magellan system. Anesthesia & Analgesia. 2013;**116**(2):491-494

[107] Webster R J, Okamura A M, Cowan N J. Toward active cannulas: Miniature snake-like surgical robots. 2006 IEEE/RSJ International Conference on Intelligent Robots and Systems. IEEE, 2006:2857-2863

[108] Shang J, Noonan D P, Payne C, et al. An articulated universal joint based flexible access robot for minimally invasive surgery. 2011 IEEE International Conference on Robotics and Automation (ICRA). IEEE. 2011:1147-1152

[109] Hempel E, Fischer H, Gumb L, et al. An MRI-compatible surgical robot for precise radiological interventions. Computer Aided Surgery. 2003;**8**(4):180-191

[110] Sutherland G R, Latour I, Greer A D, et al. An image-guided magnetic resonance-compatible surgical robot. Neurosurgery. 2008;**62**(2):286-293

[111] Xu K, Zhao J, Fu M. Development of the SJTU unfoldable robotic system (SURS) for single port laparoscopy. IEEE/ASME Transactions on Mechatronics. 2015;**20**(5):2133-2145

[112] Lee J, Kim J, Lee K K, et al. Modeling and control of robotic surgical platform for single-port access surgery. 2014 IEEE/RSJ International Conference on Intelligent Robots and Systems (IROS 2014). IEEE, 2014:3489-3495

[113] Horise Y, Matsumoto T, Ikeda H, et al. A novel locally operated master-slave robot system for single-incision laparoscopic surgery. Minimally Invasive Therapy & Allied Technologies. 2014;**23**(6):326-332

[114] Choi H, Kwak H, Kim H, et al. Surgical robot for single-incision laparoscopic surgery. IEEE Transactions on Biomedical Engineering. 2014;**61**(9):2458-2466

[115] Yuan X, Liu D, Gong M. Design and research on a shape memory alloy-actuated single-port laparoscopic surgical robot. 2014 IEEE International Conference on Mechatronics and Automation (ICMA). IEEE. 2014:1654-1658

[116] Yi B, Wang G, Li J, et al. The first clinical use of domestically produced Chinese minimally invasive surgical robot system "Micro Hand S". Surgical endoscopy. 2016;**30**(6):2649-2655

[117] Kobayashi Y, Sekiguchi Y, Noguchi T, et al. Development of a robotic system with six-degrees-of-freedom robotic tool manipulators for single-port surgery. The International Journal of Medical Robotics and Computer Assisted Surgery. 2015;**11**(2):235-246

[118] Tognarelli S, Salerno M, Tortora G, et al. A miniaturized robotic platform for natural orifice transluminal endoscopic surgery: in vivo validation[J]. Surgical endoscopy. 2015;**29**(12):3477-3484

[119] Poon C C Y, Yang H, Lau K C, et al. A bio-inspired flexible robot with hybrid actuation mechanisms for endoscopic surgery[C]//The Hamlyn Symposium on Medical Robotics. 2014: p.81

[120] Son J, Cho C N, Kim K G, et al. A novel semi-automatic snake robot for natural orifice transluminal endoscopic surgery: preclinical tests in animal and human cadaver models (with video)[J]. Surgical endoscopy. 2015;**29**(6):1643-1647

[121] Thakkar S, Awad M, Gurram K C, et al. A novel, new robotic platform for natural orifice distal pancreatectomy. Surgical innovation. 2015;**22**(3):274-282

[122] Lomanto D, Wijerathne S, Ho L K Y, et al. Flexible endoscopic robot. Minimally Invasive Therapy & Allied Technologies. 2014;**24**(1):37-44

[123] Xu K, Fu M, Zhao J. An experimental kinestatic comparison between continuum manipulators with structural variations. 2014 IEEE International Conference on Robotics and Automation (ICRA). IEEE, 2014:3258-3264

[124] Arkenbout E A, Henselmans P W J, Jelínek F, et al. A state of the art review and categorization of multi-branched instruments for NOTES and SILS. Surgical endoscopy. 2015;**29**(6):1281-1296

[125] You Z. Folding structures out of flat materials. Science. 2014;**345**(6197):623-624

[126] Liu S, Lu G, Chen Y, et al. Deformation of the Miura-ori patterned sheet. International Journal of Mechanical Sciences. 2015;**99**:130-142

[127] Liu S, Chen Y, Lu G. The rigid origami patterns for flat surface, ASME 2013 International Design Engineering Technical Conferences and Computers and Information in Engineering Conference. American Society of Mechanical Engineers. 2013;**6B**: p.1-7

[128] Liu S, Lv W, Chen Y, et al. Deployable prismatic structures with origami patterns, ASME 2014 International Design Engineering Technical Conferences and Computers and Information in Engineering Conference. American Society of Mechanical Engineers. 2014;**5B**:1-8

[129] Tolley M T, Felton S M, Miyashita S, et al. Self-folding origami: shape memory composites activated by uniform heating. Smart Materials and Structures. 2014;**23**(9):094006

[130] Miyashita S, Meeker L, Tolley M T, et al. Self-folding miniature elastic electric devices. Smart Materials and Structures. 2014;**23**(9):094005

[131] Whitney J P, Sreetharan P S, Ma K Y, et al. Pop-up book MEMS. Journal of Micromechanics and Microengineering. 2011;**21**(11):115021

[132] Hui E E, Howe R T, Rodgers M S. Single-step assembly of complex 3-D microstructures. The Thirteenth Annual International Conference on Micro Electro Mechanical Systems, 2000. MEMS 2000. IEEE, 2000:602-607

[133] Cohen A, Zhang G, Tseng F G, et al. EFAB: rapid, low-cost desktop micromachining of high aspect ratio true 3-D MEMS. Twelfth IEEE International Conference on Micro Electro Mechanical Systems, 1999. MEMS'99. IEEE. 1999:244-251

[134] Malachowski K, Jamal M, Jin Q, et al. Self-folding single cell grippers. Nano letters. 2014;**14**(7):4164-4170

[135] Shenoy V B, Gracias D H. Self-folding thin-film materials: From nanopolyhedra to graphene origami. MRS Bulletin. 2012;**37**(09):847-854

[136] Pandey S, Gultepe E, Gracias D H. Origami inspired self-assembly of patterned and reconfigurable particles. JoVE (Journal of Visualized Experiments), 2013(72): e50022-e50022.

[137] Gracias D H. Stimuli responsive self-folding using thin polymer films. Current Opinion in Chemical Engineering. 2013;**2**(1):112-119

[138] Gafford J B, Kesner S B, Wood R J, et al. Force-sensing surgical grasper enabled by pop-up book MEMS. 2013 IEEE/RSJ International Conference on Intelligent Robots and Systems (IROS). IEEE. 2013:2552-2558

[139] Salerno M, Zhang K, Menciassi A, et al. A novel 4-DOFs origami enabled, SMA actuated, robotic end-effector for minimally invasive surgery. 2014 IEEE International Conference on Robotics and Automation (ICRA). IEEE. 2014:2844-2849

[140] Majidi C. Soft robotics: a perspective—current trends and prospects for the future. Soft Robotics. 2014;**1**(1):5-11

[141] Wang Z, Chen M Z Q, Yi J. Soft robotics for engineers. HKIE Transactions. 2015;**22**(2):88-97

[142] Shepherd R F, Ilievski F, Choi W, et al. Multigait soft robot. Proceedings of the National Academy of Sciences. 2011;**108**(51):20400-20403

[143] Poon C C Y, Leung B, Chan C K W, Lau J Y W, Chiu P W Y. Design of wormlike automated robotic endoscope: dynamic interaction between endoscopic balloon and surrounding tissues. Surgical Endoscopy and Other Interventional Techniques. 2016;**30**(2):772-778

[144] Laschi C, Cianchetti M, Mazzolai B, et al. Soft robot arm inspired by the octopus. Advanced Robotics. 2012;**26**(7):709-727

[145] Wang H, Zhang R, Chen W, et al. A cable-driven soft robot surgical system for cardiothoracic endoscopic surgery: preclinical tests in animals. Surgical Endoscopy. 2016:1-7

[146] Keplinger C, Kaltenbrunner M, Arnold N, et al. Röntgen's electrode-free elastomer actuators without electromechanical pull-in instability. Proceedings of the National Academy of Sciences. 2010;**107**(10):4505-4510

[147] Onal C D, Rus D. A modular approach to soft robots. 2012 4th IEEE RAS & EMBS International Conference on Biomedical Robotics and Biomechatronics (BioRob). IEEE. 2012:1038-1045

[148] Suzumori K, Endo S, Kanda T, et al. A bending pneumatic rubber actuator realizing soft-bodied manta swimming robot. 2007 IEEE International Conference on Robotics and Automation. IEEE. 2007:4975-4980

[149] Martinez R V, Branch J L, Fish C R, et al. Robotic tentacles with three-dimensional mobility based on flexible elastomers. Advanced Materials. 2013;**25**(2):205-212

[150] Chou C P, Hannaford B. Measurement and modeling of McKibben pneumatic artificial muscles. IEEE Transactions on Robotics and Automation. 1996;**12**(1):90-102

[151] Polygerinos P, Lyne S, Wang Z, et al. Towards a soft pneumatic glove for hand rehabilitation. 2013 IEEE/RSJ International Conference on Intelligent Robots and Systems (IROS). IEEE. 2013:1512-1517

[152] Polygerinos P, Wang Z, Galloway K C, et al. Soft robotic glove for combined assistance and at-home rehabilitation. Robotics and Autonomous Systems. 2015;**73**:135-143

[153] Polygerinos P, Wang Z, Overvelde J T B, et al. Modeling of soft fiber-reinforced bending actuators. IEEE Transactions on Robotics. 2015;**31**(3):778-789

[154] Sun Y, Song Y S, Paik J. Characterization of silicone rubber based soft pneumatic actuators. 2013 IEEE/RSJ International Conference on Intelligent Robots and Systems (IROS). IEEE. 2013:4446-4453

[155] Gafford J, Ding Y, Harris A, et al. Shape deposition manufacturing of a soft, atraumatic, deployable surgical grasper. Journal of Medical Devices. 2014;**8**(3):030927

Chapter 2

Concept of Virtual Incision for Minimally Invasive Surgery

Yuki Horise, Atsushi Nishikawa, Toshikazu Kawai,
Ken Masamune and Yoshihiro Muragaki

Additional information is available at the end of the chapter

http://dx.doi.org/10.5772/intechopen.68561

Abstract

Minimally invasive surgery has been introduced to various surgical fields for its benefits such as smaller scars and less pain as compared to open surgery. Highly skilled surgical techniques are required for surgeons to conduct minimally invasive surgery with fewer ports, whereas minimally invasive surgery has a number of advantages for patients. Single-incision laparoscopic surgery (SILS), in which surgical instruments and a laparoscope are inserted through a single port, has better cosmetic results than conventional multi-incision surgery; moreover, the scar is invisible when the port is opened in navel. However, instrument collisions and visual defects often occur due to the limited space of the single opening. We propose a new surgical approach entitled "virtual incision" that enables surgeons to increase the number of openings virtually. Using our approach, we have developed two types of master-slave surgical robot systems for SILS—remote-operated and local-operated systems—which have operability close to that of multiple-incision surgery. Through evaluation of these systems, we demonstrated that the visual field and operability during virtual incision surgery are similar to those of conventional multi-incision surgery. Our surgical approach can be applied to not only single-incision surgery but also multi-incision surgery, and is very likely to improve operability.

Keywords: master-slave, robot, surgery, virtual incision

1. Introduction

In the past, curing illness was taken to be the highest priority and surgeons did not pay equal regard to patients' quality of life (QOL). Recently, however, as medical technology has progressed, equal importance can be given to both. Minimally invasive treatment has been introduced in various medical fields to reduce the patient's psychic and physical burden. For example, intravascular treatment with a catheter has been used in cerebrovascular, cardiovascular and peripheral artery surgeries, etc., and the patient's burden is drastically decreased [1]. Particle radiotherapy and advanced radiotherapy can lessen the side effects and emphasize the therapeutic effects more than conventional radiation therapy [2]. Laparoscopic surgery is a kind of minimally invasive endoscopic surgery that targets the digestive organs and requires multiple small ports to insert long surgical instruments and a cylindrical camera called a 'laparoscope'. Laparoscopic techniques bring many benefits to patients, such as smaller scars, fewer complications, and shorter hospitalization than conventional open surgery [3, 4]. To achieve less invasiveness, a new surgical method—single-incision laparoscopic surgery (SILS)—has been developed. In this method, surgical instruments, including a laparoscope, are inserted through a single opening. SILS is more cosmetic than general laparoscopic surgery since the number of ports is reduced from many to one. It is sometimes called "scar-less surgery" because the scar is invisible when the port is made at navel [5–7]. However, SILS requires high levels of skill for operating surgeons due to physical challenges such as collisions between instruments and visual defects caused by the single opening [8–10].

Minimally invasive surgery places a large burden on operating surgeons but brings great benefits for patients and preserves their QOL. To address this matter, various systems for minimally invasive surgery have been developed. The da Vinci (Intuitive Surgical Inc., CA, USA) is one of the most famous master-slave surgical robot systems in the world and can conduct laparoscopic surgery remotely with the operability close to that of open surgery by offering a 3D visual field, precise manipulation, and hand tremor cancellation. Although this robot was originally developed to support conventional multi-port laparoscopic surgery, it has also been used in single-port surgery due to its high functionality [11, 1]. Yet, it has mechanical problems such as arm collision and takes a long time to learn, even for skilled surgeons [13].

To achieve SILS that satisfies both patients and surgeons, we propose a new surgical concept called "virtual incision" and have evaluated the effectiveness of our concept by two types of master-slave robot systems for SILS.

2. Virtual incision

In SILS, the same surgical instruments and laparoscope are used as in conventional laparoscopic surgery, and the number of surgical incisions is the only difference between these operating methods. The fact that the number of surgical incisions has decreased from 3 to 1, as shown in **Figure 1(a)** and **(b)**, is one cause of the technical problems in SILS such as instruments collisions and visual defects. Thus, we propose a new surgical approach entitled "virtual incision" in which the number of ports is increased not physically but virtually. This concept provides SILS the operability close to that of multi-port surgery.

http://dx.doi.org/10.5772/intechopen.68561

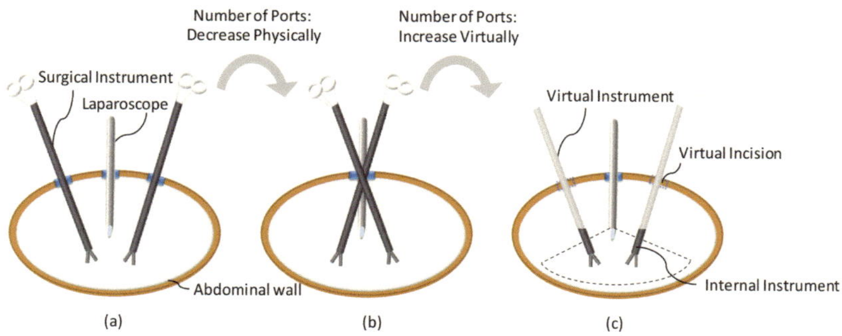

Figure 1. The number of surgical incisions required; (a) conventional multi-incision surgery, (b) single-incision laparoscopic surgery (SILS), and (c) SILS with two virtual incisions.

Making a "virtual incision" requires two conditions: (1) internal surgical instruments must behave the same as conventional laparoscopic surgery within the range of a laparoscopic view as shown in **Figures 1(c)** and (2) the operator must hold other external instruments to control the internal surgical instruments and manipulate them in the same way as conventional laparoscopic surgery. Please note that the operators normally look not at their hands but at a laparoscopic view displayed on a monitor during surgery. If these two conditions are satisfied, two virtual incisions are created on the abdominal wall in the direction of the long axis of the internal surgical instruments seen on the laparoscopic view. Therefore, surgeons can conduct SILS as if they were operating multi-port surgery even though there is only a single real port in the abdominal wall.

3. Master-slave robot systems using the concept of virtual incision

We have developed two types of master-slave robot systems using our proposed "virtual incision" approach: a remote-operated system which requires two virtual incisions, achieving three-incision surgery [14] and a local-operated system which requires one virtual incision, achieving two-incision surgery [15].

3.1. Remote-operated master-slave robot system

3.1.1. System configuration

A remote-operated master-slave robot system is mainly composed of two master instruments, two slave bending instruments and a control PC. We use two normal surgical instruments as master, which are inserted through two incisions at the tip of two holding arms set on a table as shown in **Figure 2**. A small magnetic sensor (3D-Guidance, Ascension Technology Corporation, VT, USA) is attached to each tip of the master instrument to measure the tip's position and posture. On the slave side, the two flexible instruments and a laparoscope are attached to commercial robotic arms (slave part of ZEUS surgical robot system, Computer

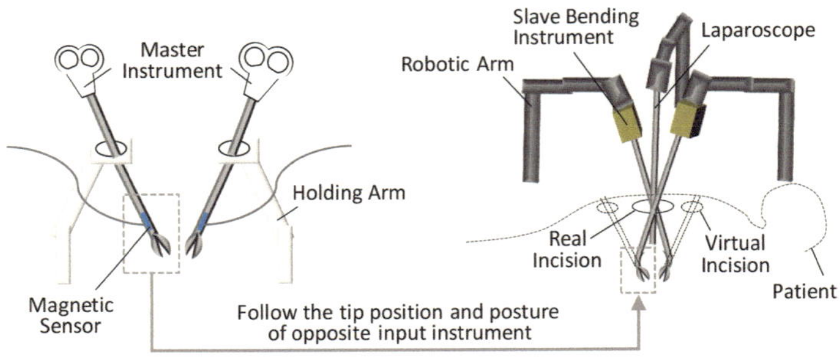

Figure 2. Correspondence between master instruments and flexible slave instruments.

Motion Inc., CA, USA), and they are all inserted into a real single opening in the patient's abdominal wall. The two slave instruments are crossed through the port and the tip of these instruments can be bent with a wire-driven mechanism. Each master instrument corresponds to the opposite slave instrument; that is, the left-side master instrument controls the slave instrument attached to right-side robotic arm; thus, the instrument tip of master behaves in the same way as that of slave in the area of the dashed square in **Figure 2**. Although the motion ratio between master and slave is adjustable (e.g., 5:1 to achieve a precise motion on the slave side), in our robot system we set the motion ratio of 1:1 to provide the surgeons the same operability as conventional laparoscopic surgery.

In this case, our proposed remote-operated master-slave robot system meets the requirements for the concept of virtual incision, then two virtual incisions are created in the abdominal wall in the long axis direction of the distal part of the slave instruments, which means that the operating surgeon can conduct SILS with three ports (one real and two virtual).

3.1.2. System evaluation

3.1.2.1. Operating conditions

Prior to system evaluation, we prepared an imitation of conventional operating conditions (multi-port and single-port surgery) by adjusting the set-up of master and slave, such as the number of ports and the shape of instruments, to compare our proposed method with conventional methods, as shown in **Figure 3**. In imitated multi-port surgery (iMPS), there are two surgical ports on the master and slave sides, and the distal part of the two slave bending instruments is set linear shape. Each master instrument corresponds to the slave instrument on the same side; thus, the same operability as in conventional multi-port surgery can be achieved. In imitated single-port surgery (iSPS), the distal parts of the two holding arms are moved close to each other on the master side, and the two slave instruments are crossed and pass through one incision, and the distal part of these is set linear shape on slave side. The master instruments correspond to the slave instruments as in the operating condition of iMPS, providing almost the same operability as conventional single-port surgery. In the proposed SILS, the configuration of and correspondence between the master and slave instruments are set as described in Section 3.1.1.

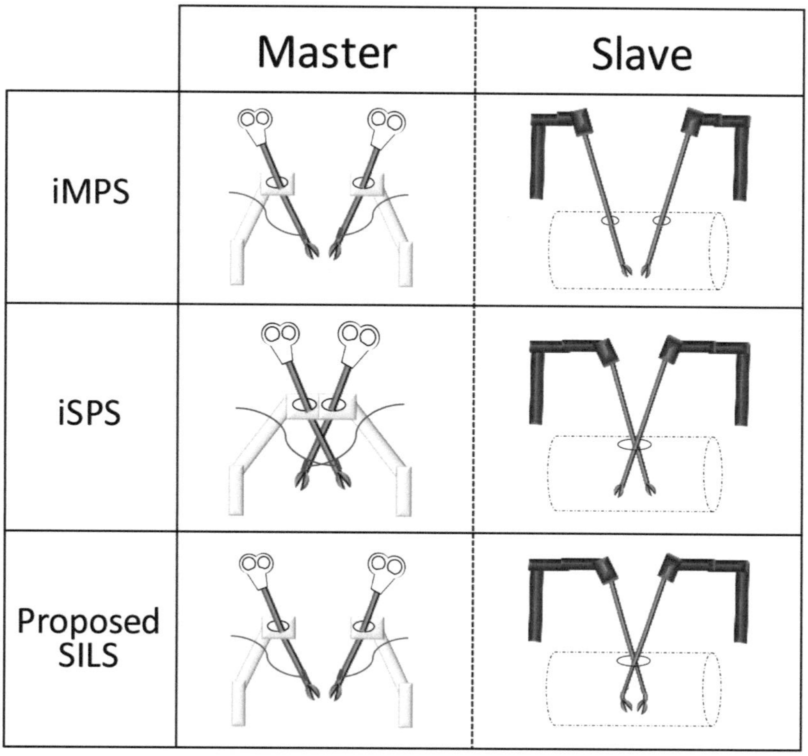

Figure 3. Master and slave configuration of three operating conditions: imitated multi-port surgery (iMPS), imitated single-port surgery (iSPS), and SILS with proposed remote-operated system (proposed SILS).

3.1.2.2. Task

In the remote-operated master-slave robot system, we conducted two experiments using (1) a simulated slave robot and (2) a real slave robot.

Experiment (1): a monitor is set behind the master, displaying an imaginary laparoscopic view as shown in **Figure 4(a)**. Using this system, four gastral surgeons conducted an object-moving task, in which the operator moves a series of balls around two boxes to each box by manipulating two master instruments, as indicated in **Figure 4(b)**. One trial is set as moving a ball 40 times, and the surgeons implemented two trials each in three operating conditions (iMPS, iSPS, and proposed SILS).

Experiment (2): two robotic arms are mounted on a surgical bed and two bending slave instruments are attached to each arm. A box is placed on the surgical bed, and the two slave instruments and a laparoscope are inserted into the box, as shown in **Figure 5**. A monitor is set as well as in Experiment (1) and a real laparoscopic view is displayed. Under this experimental environment, another gastral surgeon conducted an object-touching task, in which the operator touches a cylindroid object in the box using the two slave instruments by manipulating the two master instruments. He conducted seven trials in each of the two operating conditions (iSPS and proposed SILS).

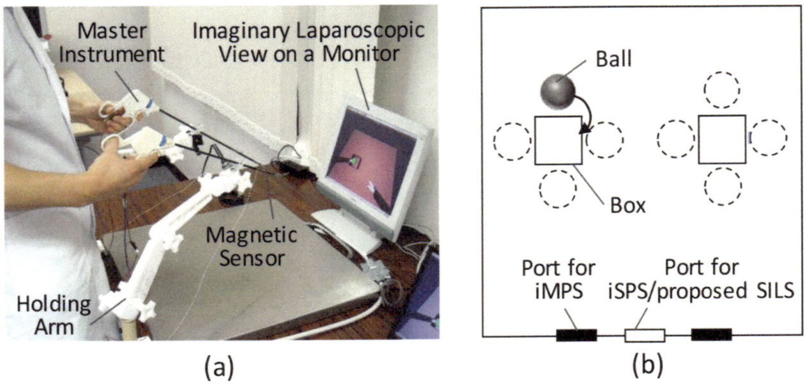

Figure 4. Object-moving task: (a) experimental set-up during the task in iMPS and (b) task detail; black port for iMPS and white port for iSPS and proposed SILS.

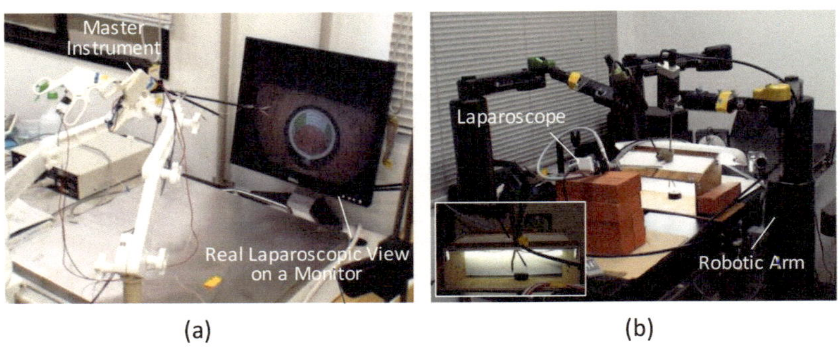

Figure 5. Object-touching task during the task in iSPS: (a) master set-up and (b) slave set-up.

3.1.2.3. Result

Experiment (1): average object-moving times among the four surgeons for each trial were 251 ± 61 s (iMPS), 310 ± 52 s (iSPS), and 247 ± 67 s (proposed SILS) in trial 1, and 182 ± 29 s (iMPS), 248 ± 77 s (iSPS), and 178 ± 15 s (proposed SILS) in trial 2. There was no significant difference between iMPS and proposed SILS ($p = 0.23 > 0.1$), and a significant difference between iSPS and proposed SILS was confirmed ($p = 0.011 < 0.05$).

Experiment (2): average object-touching times were 25 ± 7 s (iSPS) and 11 ± 4 s (proposed SILS).

3.2. Local-operated master-slave robot system

3.2.1. System configuration

A local-operated master-slave robot consists of a master device, a slave robotic instrument, and a control PC. The master device is placed above the patient's abdominal wall and can be fixed to a surgical table using a conventional passive grasping holder. The slave robotic instrument is also attached to the surgical table with another grasping arm, and the robotic instrument can change between the two states of straight or bent; it is straight when passing through an incision and then is bent and set in alignment with the master device in the body as indicated in **Figure 6(a)**. The master device and slave robotic instrument have five-DOF motion as in conventional laparoscopic surgery, including a grip/tip opening and closing (OC), spherical motion around the base point (pitch and yaw), axial rotation (roll), and extraction and contraction (EC) along the longitudinal axis. The posture and tip opening and closing of the slave robotic instrument have a point-symmetrical relationship with those of the master device, and the length of the slave instrument is inversely proportional to that of the master device. The master device and slave robotic instrument are moved simultaneously through the abdominal wall with a motion ratio of 1:1, thus they behave like a normal commercial surgical instrument inserted into a port. Our proposed system could be used as a substitute for the left-sided surgical instrument in SILS, especially in single-incision laparoscopic cholecystectomy as shown in **Figure 6(b)** since the left-sided surgical instrument is often used for grasping tissue.

In this case, our proposed local-operated master-slave robot system meets the requirements of virtual incision, and one virtual incision is created in the abdominal wall between the master device and the slave robotic instrument as illustrated in **Figure 6(a)**. This means that the operating surgeon can conduct SILS with two ports (one real and one virtual).

3.2.2. System evaluation

3.2.2.1. Task

In the local-operated master-slave robot system, we conducted two experiments: (1) a basic experiment and (2) an *ex vivo* experiment using a porcine liver.

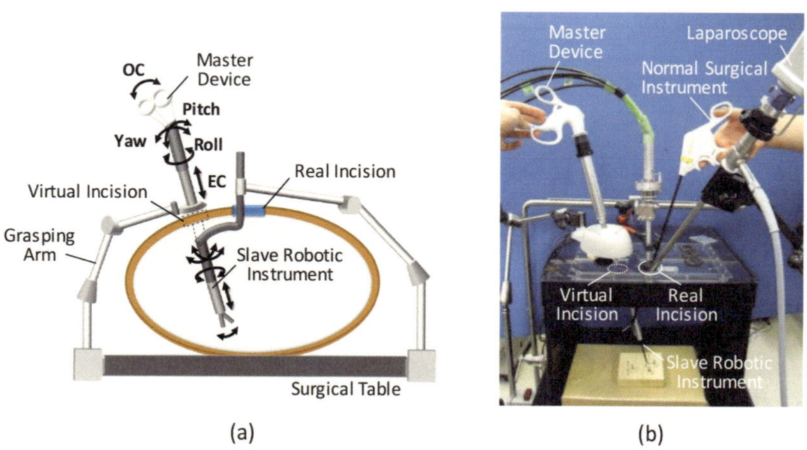

Figure 6. Local-operated master-slave robot system: (a) concept of proposed system and (b) surgical instrumental set-up for single-incision laparoscopic cholecystectomy.

Experiment (1): a training box is placed on a surgical table, and a master device is set above the box as a left-side surgical instrument as shown in **Figure 7(a)**. A right-sided normal surgical instrument, a laparoscope, and a slave robotic instrument are all inserted into the single real opening. The operator moves a series of rings placed on a peg board in the box from left to right using the surgical instruments with both hands while observing the laparoscopic view on a monitor. The ring is lifted from the left-side peg by the left-side instrument in the laparoscopic view, transferred to the opposite-sided instrument, and placed on the right-side peg, as illustrated in **Figure 7(b)**. To evaluate the operability of our proposed SILS, we prepared two conventional SILS operating conditions—SILSc and SILSp—in which two normal surgical instruments are inserted through a single port and are manipulated with crossed and parallel set-ups, respectively. One trial is set as moving three rings three times, and three gastral surgeons (Surgeon A, B, and C) conducted three trials each under three operating conditions (SILSc, SILSp, and proposed SILS).

Experiment (2): the operating condition of proposed SILS is set as in Experiment (1), and a normal resection instrument is used as a right-sided instrument. As shown in **Figure 7(c)**, a porcine liver with a gallbladder is fixed to a board in the training box, and the cystic duct has previously been clipped. In this experiment, we simulate a single-port laparoscopic cholecystectomy, and an operator separates the gallbladder from the liver by manipulating the surgical instruments with both hands while watching the laparoscopic view.

3.2.2.2. Result

Experiment (1): average switching time between the instruments and standard deviation of Surgeon A were 11.65 ± 14.97 s (SILSc), 21.64 ± 33.85 s (SILSp), and 7.03 ± 2.49 s (proposed SILS); those of Surgeon B were 6.82 ± 4.51 s (SILSc), 13.82 ± 29.40 s (SILSp), and 8.48 ± 3.58 s (proposed SILS); and those of Surgeon C were 10.04 ± 5.07 s (SILSc), 35.82 ± 34.66 s (SILSp),

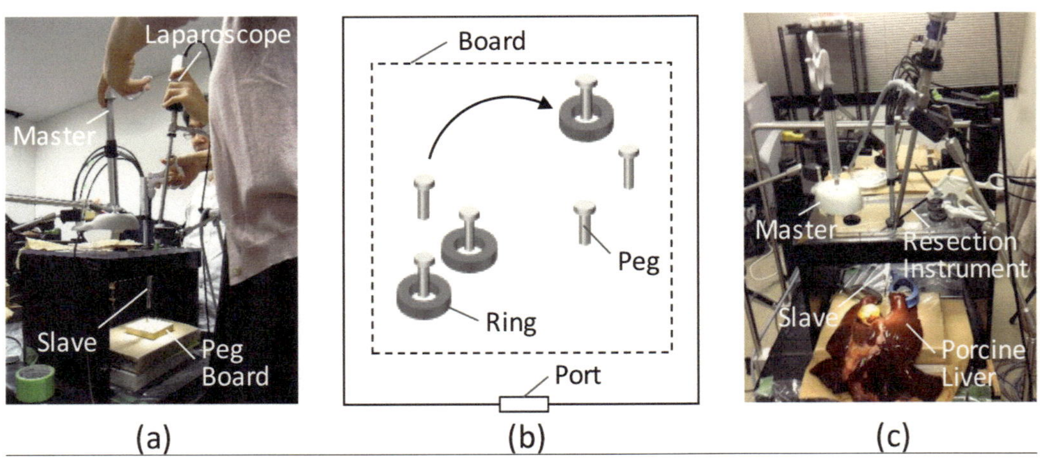

Figure 7. Evaluation experiment of local-operated master-slave robot system: (a) object-moving task in the operating condition of our proposed SILS, (b) peg board detail, and (c) *ex vivo* experimental set-up in single-port laparoscopic cholecystectomy.

and 6.48 ± 2.25 s (proposed SILS). The average switching times among the three surgeons were 9.50 ± 9.70 s (SILSc), 22.25 ± 33.57 s (SILSp), and 7.23 ± 2.87 s (proposed SILS).

The number of collisions between the instruments in the box during the task of Surgeon A was 7 times (SILSc), 15 times (SILSp), and 0 times (proposed SILS); those of Surgeon B were 3 times (SILSc), 9 times (SILSp), and 1 time (proposed SILS); and those of Surgeon C were 1 time (SILSc), 18 times (SILSp), and 0 times (proposed SILS). The average number of collisions of all surgeons was 3.67 times (SILSc), 14 times (SILSp), and 0.33 times (proposed SILS). Meanwhile, the collision between the instrument and laparoscope occurred frequently outside the box under two conventional SILS operating conditions; however, the master device did not conflict with the laparoscope or the other instrument in our proposed SILS.

Experiment (2): the operator separated the gallbladder from the liver without instrument collision, grasping it using the slave robotic instrument by manipulating the master device with the left hand and cutting it using the resection instrument with the right hand.

Laparoscopic views during Experiments (1) and (2) are presented in **Figure 8**.

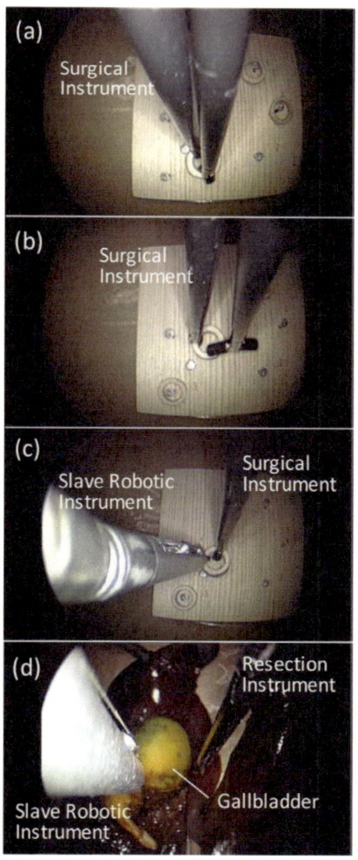

Figure 8. Laparoscopic view during the task; object-moving task in the operating condition of (a) SILSc, (b) SILSp, (c) proposed SILS, and (d) *ex vivo* cholecystectomy using our local-operated master-slave robot system.

4. Discussion

As for the remote-operated master-slave robot system, from Experiment (1), we confirmed that the average object-moving time and standard deviation in the proposed SILS were significantly shorter than in iSPS, representing conventional single-port surgery, and were almost the same as those in iMPS, representing conventional multi-port surgery. From the subjective evaluations, all surgeons commented that the operability in the proposed SILS was little differentiated from that in iMPS. In Experiment (2), the object-touching task in the proposed SILS was accomplished in shorter time than iSPS, and there was less interference between the instruments in the proposed SILS than in iSPS. These results indicate that our proposed remote-operated system could improve the operability of SILS and achieve operability close to conventional multi-port surgery by adding two virtual incisions.

In regard to the local-operated master-slave robot system, in Experiment (1), the average switching time and standard deviation for proposed SILS were shorter than those for two conventional SILS operating conditions (SILSc and SILSp). Instances of interference between the surgical instruments were drastically decreased in the proposed SILS as compared with SILSc and SILSp. Focusing on the left-side surgical instrument, its location in the laparoscopic view in our proposed SILS differed from that of conventional SILS methods, as shown in **Figure 8(a)–(c)**. The left-side and right-side instruments were inserted from the same direction on the laparoscopic view and were close together in SILSc and SILSp, and the left-side instrument was inserted from left side on the laparoscopic view as in conventional multi-port laparoscopic surgery in proposed SILS. The subjective evaluations from the surgeons also indicated that they were able to conduct SILS more easily using our proposed system than conventional SILS. In Experiment (2), the operator performed single-port laparoscopic cholecystectomy without any problems, and a laparoscopic view close to that in multi-port surgery was observed as shown in **Figure 8(d)**. The operator also reported that it was like performing laparoscopic surgery though there was only one port. From these results, we confirmed that the local-operated master-slave robot system could improve the operability of SILS, providing greater efficiency and stability, and offering the operability and a surgical view similar to those of multi-port laparoscopic surgery by introducing one virtual incision.

From the above results of the two master-slave robot systems, we have confirmed the usefulness of our proposed new "virtual incision" surgical approach, since it brings the operability and surgical view similar to multi-port laparoscopic surgery even though only one port is opened in the abdominal wall. The operational procedure of our proposed surgical robot systems is close to the conventional surgical method, and thus does not require extra training for surgeons who are used to conventional multi-port laparoscopic surgery.

Although the initial motivation for proposing the "virtual incision" concept is to overcome the difficulties of SILS, such as instrument collision and visual defects, this concept can be used not only for single-port surgery but also for conventional multi-port surgery. For example, three-port surgery could be conducted with two real openings by adding one "virtual incision". In thoracoscopic surgery, the surgical port position is limited by the alignment of the limb, and is not as flexible as in laparoscopic surgery. By introducing the concept of "virtual incision" to thoracoscopic surgery, it is expected that surgeons can make a surgical port at an appropriate position without worrying about the limb, and the operating area can expand with fewer ports. We plan to develop a surgical device introducing our concept for thoracoscopic surgery [16].

Needlescopic surgery, another form of minimally invasive surgery, has been utilized in various surgical fields and brings cosmetic benefits for patients since it uses surgical instruments with very small diameter [17, 18]; however, it also has problems as it is difficult to keep the stiffness of the surgical instrument while including adequate functionality in a small-diameter instrument. We believe that our new surgical concept of "virtual incision" is one way to achieve more minimally invasive surgery by reducing the number of real ports through a virtual incision.

As for master-slave surgical robot systems, many surgical robot systems have been developed to support surgery, but a master can control only a slave of the same system. For future robotic surgery, we think that collaboration among various master-slave surgical robot systems will become increasingly important. Operating surgeons have different preferences for the master controller of a system, and it is difficult to decide on one master. Robotic surgery could become more flexible if multiple masters can control the same slave or one slave can be easily exchanged for another. We are also working on another project in which multi-master and multi-slave options could be selected flexibly using an industrial middleware ORiN for a next-generation surgical-assisted robot system [19].

5. Conclusion

Minimally invasive approaches such as reduced port surgery have been introduced to various surgical fields to achieve more patient-friendly surgery. Although single-incision laparoscopic surgery (SILS) has cosmetic benefits for patients, such as less pain, smaller scar, and shorter hospitalization than conventional laparoscopic surgery due to its single opening, it has technical problems that increase the surgeon's burden. Therefore, we have proposed a new surgical approach called a "virtual incision", and have applied it to two types of master-slave robot systems for assisting SILS. From the evaluation results of the two master-slave robot systems, we have confirmed the effectiveness of "virtual incision" because the operability of SILS was improved by increasing the number of ports virtually. In this chapter, we have focused on SILS; however, our surgical approach could also be useful to multi-port surgery. We believe that the concept of "virtual incision" is a promising surgical approach that preserves QOL for both patients and surgeons.

Acknowledgements

This research was partly supported by MEXT KAKENHI Grant Number 15K16338 and 16H01 859.

Author details

Yuki Horise[1], Atsushi Nishikawa[2*], Toshikazu Kawai[3], Ken Masamune[1] and Yoshihiro Muragaki[1]

*Address all correspondence to: nishikawa@shinshu-u.ac.jp

1 Faculty of Advanced Techno-Surgery, Tokyo Women's Medical University, Tokyo, Japan

2 Faculty of Textile Science and Technology, Shinshu University, Ueda, Japan

3 Faculty of Engineering, Osaka Institute of Technology, Osaka, Japan

References

[1] Lange R, Bleiziffer S, Mazzitelli D, Elhmidi Y, Opitz A, Krane M, Deutsch M-A, Ruge H, Brockmann G, Voss B, Schreiber C, Tassani P, Piazza N. Improvements in transcatheter aortic valve implantation outcomes in lower surgical risk patients: A glimpse into the future. Journal of the American College of Cardiology. 2012;**59**(3):280-287

[2] Grutters JPC, Kessels AGH, Pijls-Johannesma M, De Ruysscher D, Joore MA, Lambin P. Comparison of the effectiveness of radiotherapy with photons, protons and carbon-ions for non-small cell lung cancer: A meta-analysis. Radiotherapy and Oncology. 2010;**95**(1):32-40

[3] Lacy AM, García-Valdecasas JC, Delgado S, Castells A, Taurá P, Piqué JM, Visa J. Laparoscopy-assisted colectomy versus open colectomy for treatment of non-metastatic colon cancer: A randomized trial. The Lancet. 2002;**359**(9325):2224-2229

[4] Prasad A, Mukherjee KA, Kaul S, Kaur M. Postoperative pain after cholecystectomy: Conventional laparoscopy versus single-incision laparoscopic surgery. Minimally Access Surgery. 2011;**7**(1):24-27

[5] Desai MM, Rao PP, Aron M, Pascal-Haber G, Desai MR, Mishra S, Kaouk JH, Gill IS. Scarless single port transumbilical nephrectomy and pyeloplasty: First clinical report. Journal Compilation BJU International. 2008;**101**(1):83-88

[6] Ross S, Rosemurgy A, Albrink M, Choung E, Dapri G, Gallagher S, Hernandez J, Horgan S, Kelley W, Kia M, Marks J, Martinez J, Mintz Y, Oleynikov D, Pryor A, Rattner D, Rivas H, Roberts K, Rubach E, Schwaitzberg S, Swanstrom L, Sweeney J, Wilson E, Zemon H, Zundel N. Consensus statement of the consortium for LESS cholecystectomy. Surgical Endoscopy. 2012;**26**(10):2711-2766

[7] Autorino R, Kaouk JH, Stolzenburg J-U, Gill IS, Mottrie A, Tewari A, Cadeddu JA. Current status and future directions of robotic Single-Site surgery: A systematic review. European Urology. 2013;**63**(2):266-280

[8] Romanelli JR, Earle DB. Single-port laparoscopic surgery: An overview. Surgical Endoscopy. 2009;**23**(7):1419-1427

[9] Mutter D, Callari C, Diana M, Dallemagne B, Leroy J, Marescaux J. Single port laparoscopic cholecystectomy: Which technique, which surgeon, for which patient? A study of the implementation in a teaching hospital. Journal of Hepato-Biliary-Pancreatic Sciences. 2011;**18**(3):453-457

[10] Ma J, Cassera MA, Spaun GO, Hammill CW, Hansen PD, Aliabadi-Wahle S. Randomized controlled trial comparing single-port laparoscopic cholecystectomy and four-port laparoscopic cholecystectomy. Annals of Surgery. 2011;**254**(1):22-27

[11] Erbella Jr J, Bunch GM. Single-incision laparoscopic cholecystectomy: The first 100 outpatients. Surgical Endoscopy. 2010;**24:**1958-1961

[12] Kroh M, El-Hayek K, Rosenblatt S, Chand B, Escobar P, Kaouk J, Chalikonda S. First human surgery with a novel single-port robotic system: Cholecystectomy using the da Vinci Single-Site platform. Surgical Endoscopy. 2011;**25:**3566-3573

[13] Park S-S, Kim M-C, Park MS, Hyung WJ. Rapid adaptation of robotic gastrectomy for gastric cancer by experienced laparoscopic surgeons. Surgical Endoscopy. 2012;**26**:60-67

[14] Horise Y, Nishikawa A, Sekimoto M, Kitanaka Y, Miyoshi N, Takiguchi S, Doki Y, Mori M, Miyazaki F. Development and evaluation of a master-slave robot system for single-incision laparoscopic surgery. International Journal of Computer Assisted Radiology and Surgery. 2012;**7**(2):289-296

[15] Horise Y, Matsumoto T, Ikeda H, Nakamura Y, Yamasaki M, Sawada G, Tsukao Y, Nakahara Y, Yamamoto M, Takiguchi S, Doki Y, Mori M, Miyazaki F, Sekimoto M, Kawai T, Nishikawa A. A novel locally operated master-slave robot system for single-incision laparoscopic surgery. Minimally Invasive Therapy & Allied Technologies. 2014;**23**(6):326-332

[16] Horise Y, Nishikawa A, Kawai T, Kanzaki M, Masamune K, Muragaki Y. Endoscopic surgical support system with a concept of "Virtual Incision". Journal of Japan Society of Computer Aided Surgery. 2015;**17**(3):277-278

[17] Cheah WK, Lenzi JE, So JBY, Kum CK, Goh PMY. Randomized trial of needlescopic versus laparoscopic cholecystectomy. British Journal of Surgery. 2001;**88**(1):45-47

[18] Mamazza J, Schlachta CM, Seshadri PA, Cadeddu MO, Poulin EC. Needlescopic surgery. Surgical Endoscopy. 2001;**15**(10):1205-1212

[19] Okamoto J, Horise Y, Masamune K, Iseki H, Muragaki Y. Development of a prototype model of "Hyper SCOT (Smart Cyber Operating Theater)". International Journal of Computer Assisted Radiology and Surgery. 2016;**11**(Supplement 1):S163

Section 2

Applications in Robotic Surgery

Chapter 3

The Surgical Robot: Applications and Advantages in General Surgery

Rodolfo José Oviedo Barrera

Additional information is available at the end of the chapter

http://dx.doi.org/10.5772/intechopen.68864

Abstract

The field of General Surgery with its multiple sub-specialties has experienced the progression of minimally invasive procedures performed with the robotic technology since the last decade. The robotic applications are extensive and have contributed to the enrichment of the surgical sub-specialties based on advantages such as increased surgeon control and autonomy, superior instrument dexterity and tissue handling, improved three-dimensional visualization, wristed articulation, all of this despite the lack of haptic feedback. The sub-specialties of Colorectal, Hepatobiliary and Pancreatic, Gastric Oncologic, Bariatric, Foregut, Pediatric, Endocrine, and Hernia Surgery, in addition to General Surgery as the principal specialty, have produced several high-quality randomized controlled trials, meta-analyses, prospective and retrospective series which have established, in many instances, superior results to those of laparoscopy, and at least non-inferior outcomes over the years. From the first pioneer single-surgeon experiences around the world to the most recent large trials, including the first Robotic General Surgery case series in an American community hospital not classified as a tertiary referral center, patients continue to benefit from this technology as surgeons engage in overcoming their learning curve and training their teams, involving their hospital administrators and working with the industry to perfect their techniques for the sake of their patients.

Keywords: surgery, general, robotic, colorectal, hepatobiliary, pancreatic, gastric, oncology, bariatric, foregut, hernia, pediatric, endocrine, learning, curve, technology

1. Origins of a revolution

For general surgeons, it should be easy to define their specialty. For the public, however, the term "General Surgery" may carry the erroneous implication of a lack of specialization, a deficiency in expertise, or even a certain weakness of purpose. To define what General Surgery is and stands for, it should be established that it is the mother of all surgical subspecialties, a means to save lives from traumatic experiences, to cure cancer, to offer palliation and improve quality of life, to remove organs that suffer from overwhelming infections, and to reconstruct the body's tissues and organ systems. To restore anatomy and physiology, while life acquires a higher quality, that is the ultimate purpose of General Surgery.

In order to discuss the robotic revolution in General Surgery, it is necessary to establish that this was the last surgical specialty that adopted the robotic technology, first with hesitation. However, to this day, the progress of the robotic technology applications in this field is palpable and replicated by numerous surgeons in the academic and the private practice environment around the world.

The concept of robotics applied to perform an operation has been explored extensively since the end of the twentieth century and the beginning of the twenty-first century, including the development of robotic platforms such as the AESOP/Hermes, the Zeus, and the da Vinci systems [1, 2]. This effort on behalf of multiple companies and research centers, including NASA, led to the development of the telerobotic technology necessary for different specialties to adopt it in order to carry out surgery in a minimally invasive fashion while overcoming some of the obstacles that laparoscopic surgery introduced at the end of the 1980s [3].

However, in spite of the major achievements that robotics in General Surgery has witnessed thanks to its ability to enable minimally invasive surgeons to overcome some of their limitations, even up to a few years ago and to this date there is opposition to the use of robotic surgery. A typical reason that is often quoted is the apparent usefulness of the robot only for certain subspecialties such as colorectal surgery given the limited working space in the pelvis and the challenge posed by traditional laparoscopic instrumentation. This is in addition to the financial burden that the application of robotic surgery carries with it when the conscientious use of only the necessary instruments is not a priority [4].

Despite the reluctance to the widespread adoption of robotics in General Surgery, many surgeons around the world have already been responsible for the advancement of surgery in their fields in all of the disciplines or subspecialties that will be presented in this chapter, such as colorectal, hepatobiliary and pancreatic, gastric oncology, bariatric and antireflux, pediatric, endocrine, and hernia/abdominal wall reconstruction surgery. The purpose of this chapter is to describe these achievements in an objective way, so that the idea that the surgical robot should only be used for colorectal surgery or complex foregut surgery may be challenged and, furthermore, so that this author's passion for robotic surgery may be shared with the international surgical and scientific community for the sake of the patients' well-being.

2. Where engineering meets medicine

Robotic General Surgery has advanced at an accelerated pace since the late 2000s, although early studies as far back as 2004 expressed concerns that the field was in its infancy an lacked the necessary data to substantiate its widespread use and its safe application. Nevertheless, even at that early point in the history of robotics in the largest surgical specialty, the multiple advantages of robotic surgery were recognized and described as the ability to have wristed instrumentation with more degrees of motion than the human hand is capable of acquiring, superior visualization with three-dimensional capability and with surgeon control of the camera, the presence of more than two arms to execute tasks, which facilitate the creation of anastomoses with superior dexterity, along with more advanced ergonomics than what can be provided with conventional laparoscopic instruments. On the other hand, the disadvantages were not technical except for the lack of haptic (tactile) feedback for the surgeon. The other disadvantages had to do with systems and processes not related to the technical aspects of an operation, such as the cost of instrumentation, the cost associated with purchasing the technology, the intensive nature of training for the surgeon and the team, and the apparent unproven benefit in all branches of General Surgery, at least as it was seen at that time [5].

Based on expert surgeons' personal experience, however, the most important advantage offered by the surgical platform is the ability to offer them total control of the procedure without the need to depend on someone else to operate the endoscope, or retract, or assist in a manner that would be crucial with conventional laparoscopy. While complex robotic surgery still requires a first assistant, the assistant's role has evolved because the surgeon has total control of three arms at the same time along with the camera, all of which enables the operator to achieve the goals in a manner that is closer to open surgery, at least closer than ever before.

From an engineering perspective, it is essential for surgeons to understand the concept of telerobotics and the categorization of robotics in General Surgery as a short-distance system consisting of a "master" component operated by the surgeon, and a "slave" executor which carries out the tasks performed by the "master" platform in real time. By definition, this is not an autonomous or semiautonomous technology, which is an important point to clarify, since it means that the surgical robot does not have the capacity to operate itself for a reason: it maintains the surgeon's total control of the procedure enabled by a computer interface that facilitates the execution of the operation. This is the definition of a "tele-operator" system (see **Figures 1** and **2**).

As mentioned earlier, the da Vinci system (by Intuitive Surgical, Inc., Sunnyvale, CA, USA) was developed while building upon the lessons learned from its predecessors such as the AESOP and the Zeus platforms. It consists of an ergonomic console unit ("master interface") that includes a display system, the surgeon's user interface and the controller, and a second unit that includes the endo-wristed instruments and the endoscopic camera that execute the tasks as the "slave manipulator." Its application in all fields of General Surgery has been documented extensively, although, initially, it was created to satisfy the minimally invasive needs of cardiothoracic surgeons and urologists, and later on, gynecologists [6].

Even in 2008, at the time when widespread adoption of the robotic interface was beginning to take place among general surgeons, the disadvantage of lack of haptic feedback was studied, with results being consistent with the absence of consensus among the surgical community regarding its essential value to perform an operation. In fact, although the ability to have haptic feedback has been generally considered a useful feature of laparoscopic surgery, its

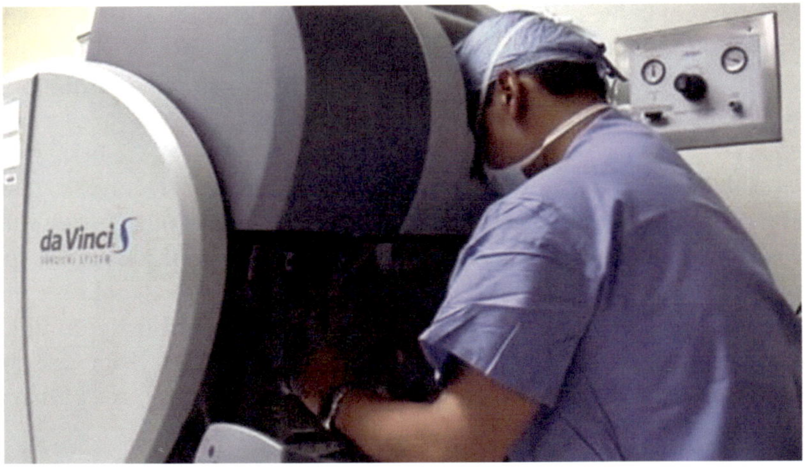

Figure 1. Surgeon operating at the ergonomic console unit ("master interface"). The user interface allows for "endowrist" articulation of instruments, with seven degrees of freedom for motion.

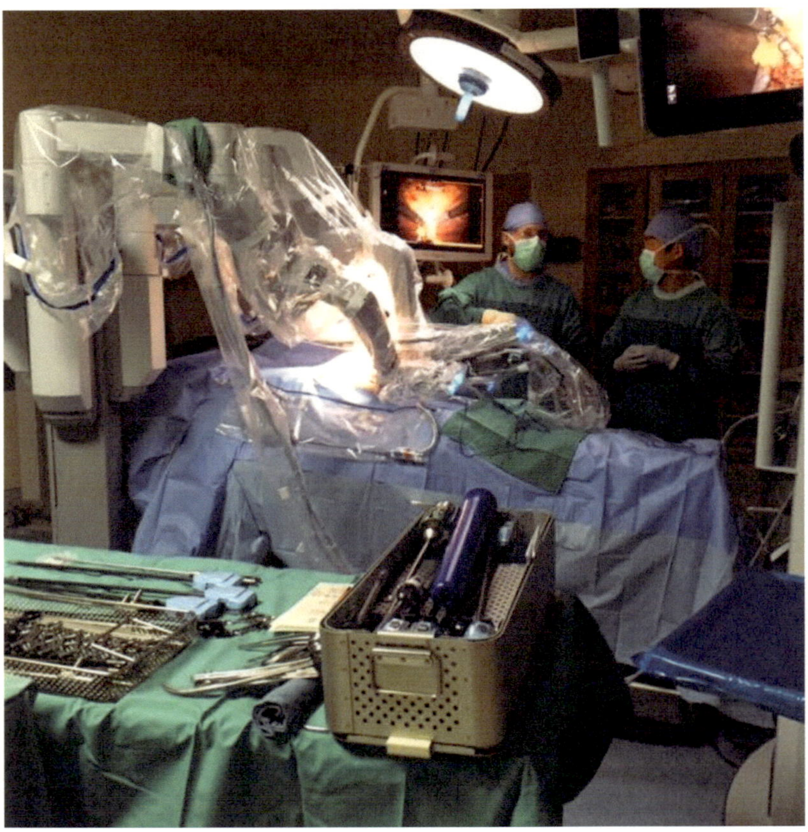

Figure 2. da Vinci system ("slave manipulator") with robotic arms already docked and executing the tasks with the surgeon in control, with the surgical team at the bedside.

absence in robotic surgery can be overcome by the superior visibility offered by the surgeon-controlled three-dimensional endoscope and visual cues when tissue tension is carefully observed [7].

From a technical perspective, the robotic technology enables the surgeon to overcome the challenges that traditional laparoscopic surgery offers, as it has been described. However, an important aspect of this ability to improve the surgeon's skill level can be seen when it is used for practice purposes, both by expert surgeons and by inexperienced surgeons who are trying to develop their skill set to offer the multiple benefits of minimally invasive surgery to their patients. Reductions in errors have been noticed when such practice tasks are undertaken for the purpose of quality and self-improvement [8].

What is impressive, considering the early period when another study was presented at an important surgical society meeting, a successful robotic surgery training program can be implemented, with reproducible and reliable results, as long as the will and determination exist to apply the benefits of robotic surgery and transform them into palpable outcomes with the highest ethical and quality standards in an academic institution [9].

3. Colorectal surgery: the subspecialty that paved the way

The cost of robotic surgery has always been an element of strong criticism used against its adoption in multiple surgical subspecialties, including the pioneer, colorectal surgery. However, even in those well-conducted studies, the benefits of robotic surgery have been noted without a doubt, such as better outcomes in left colectomies, particularly when approaching the rectum when compared to even the most sophisticated 3-D laparoscopic systems [10]. As early as 2013, several manuscripts in the field of robotic colorectal resections were analyzed and the conclusions suggested that robotic surgery would continue to advance and overcome its own weaknesses, with improved outcomes comparable to those of conventional laparoscopy [11].

A more recent review of the colorectal literature, although not in favor of robotic surgery, acknowledges the established advantages over laparoscopic colorectal resections that have been reported by multiple series including decreased blood loss, decreased length of hospitalization, faster return of bowel function and, what is more interesting, a lower rate of conversion to open surgery [12]. Similar conclusions have been drawn from an extensive meta-analysis in 2015 comparing robotic versus laparoscopic colorectal resections, which also pointed out a lower incidence of peri-operative complications and surgical site infections [13].

However, perhaps more significant progress could be achieved once the robotic surgery is not compared to laparoscopic surgery. Conclusions from another manuscript in a prestigious journal have suggested that although it is feasible and safe to perform robotic surgery for sigmoid colon resections for cancer, it offers no real advantage over laparoscopic surgery in terms of oncologic outcomes [14]. Even another publication reported on the feasibility and safety of robotic transverse colon resection for cancer, too [15]. This is an important shifting paradigm from the tradition of comparing once technology against the other, which is sometimes a reason for many surgeons to hesitate when it comes to deciding to adopt robotics as part of their practice.

More published results from well-done meta-analyses support the superiority of robotic surgery in colorectal resections for oncologic purposes, with the same conclusions already mentioned in terms of blood loss, safety, the length of stay, the return of bowel function, lower estimated blood loss, and conversion rates [16, 17]. On the other hand, the efficiency of the robotic platform can be seen when an oncologic resection is performed, as the number of lymph nodes is comparable to that obtained with laparoscopy by the most experienced surgeons [18] Returning to the issue of cost, robotic segmental colon resections have been associated with increased operative time, perhaps due to the surgeons' learning curve, in addition to overuse of non-essential instrumentation [19].

More specifically on the subject of rectal cancer, robotic surgery has been found highly efficacious and comparable to open surgery, with similar oncologic outcomes, lymph node yields, free margins, disease-free survival, and rate of complications. The length of the operation is greater, but this is something where improvement can be seen with increased volumes [20].

Regarding rectal cancer and the need for total mesorectal excision, which has been a topic of continuous discussion in the literature over the years, the robotic platform has been found to offer superior results for mid and low rectal cancer resections, where the quality of the TME specimen has been documented to be more advanced than its laparoscopic counterpart (see **Figure 3**). Moreover, conversion rates to the open approach have been determined to be lower thanks to the robotic platform advantages explained in detail before [21, 22].

Another aspect of robotic rectal resections for cancer is the facilitation of an oncologic resection with the Firefly™ technology, which has proven very helpful during low ligation of the inferior mesenteric artery pedicle. The ability to perform a precise lymphadenectomy around the IMA is invaluable, all of which is made possible with the robot's multiple benefits when it comes to retroperitoneal and pelvic dissection [23]. The most challenging lymphadenectomy, however, at least in the colorectal surgery arena, corresponds to the total mesorectal excision technique. It is under difficult circumstances of a narrow male pelvis, or a female pelvis that has been previously subjected to radiation therapy, where the fibrosis and desmoplastic reaction from a neoplastic process require the surgeon's maximum level of proficiency for the sake of a safe, efficient oncologic resection. The robotic technology enables the surgeon to achieve excellent results where laparoscopic surgery has failed to deliver in the past [24, 25]. Interestingly, it has been determined that the learning curve for robotic low anterior resection (including total mesorectal excision) is similar and not longer than the learning curve for the laparoscopic technique, which argues against the idea that it would be more difficult to learn to perform such a demanding and challenging procedure with the robot as opposed to doing it laparoscopically. This is not to say that robotic LAR and TME are not highly technical procedures that require a remarkable level of skill to be carried out well, but they can be learned [26–28].

On a separate subject, robotic surgery in the colorectal field has also been extremely useful when it comes to benign disease, which is sometimes more complex than procedures done for neoplastic processes. The perfect example is diverticular pathology with colovesical fistula resection and repair. A study has compared the laparoscopic to the robotic technique. The

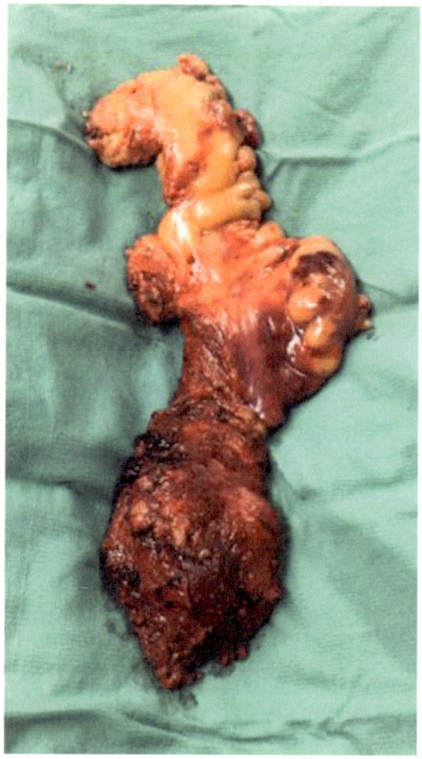

Figure 3. Robotic ultra low anterior resection and total mesorectal excision specimen for rectal adenocarcinoma.

remarkable observation of this series was the fact that the robotic group did not experience any conversions to open surgeries, or any ureteral injuries. The same was not true of the laparoscopic arm [29]. Along the lines of benign disease, rectal prolapse, and robotic rectopexy have been studied and compared to the laparoscopic approach, with the conclusion that both methods to deal with it are effective, although more data are needed to establish any superiority of the robotic technique, such as a randomized controlled trial. Be that as it may, the important aspect of this study is the fact that the surgical robot can be very effective when it comes to benign colorectal disease and its use can be safely expanded to treat conditions that would normally be dealt with open surgery [30, 31].

4. Hepatobiliary and pancreatic surgery: nothing is impossible

Without a doubt, the field of hepatobiliary and pancreatic surgery is highly regarded as one of the most complex and technically demanding subspecialties within General Surgery. In fact, a pancreaticoduodenectomy is considered by most surgeons as the most difficult operation in the world, second perhaps to a liver transplant. What seemed impossible years ago has become a reality with arduous determination and the process of trial and error, where numerous experts have advanced this field to the realm of the minimally invasive and have turned operations that would typically be unthinkable or impractical with laparoscopy into reproducible robotic procedures whose results will be analyzed here.

In 2013, the largest retrospective series of robotic pancreatic resections was published, which comprised 250 operations ranging from pancreaticoduodenectomies, to central, distal, and total pancreatectomies. This impressive series demonstrated the feasibility of oncologic and benign disease resections with a low conversion rate [32]. A more modest series of 12 patients reported the same year drew similar conclusions while emphasizing the importance of clinical judgment at all times, which serves as a reminder that the robotic technology is just a tool at the service of the surgeon, who is ultimately responsible for the outcome of any operation [33].

A comprehensive literature review the next year also reached these conclusions and warned the surgical community that the series that were examined had their origin in academic centers where the experts in their field performed these procedures, all within hospital systems that had the human and technical capability to deal with the complications that are known to be inherent to pancreatic surgery [34]. When different series are reviewed, the most important advantage from the robotic technology that is strongly applied to pancreatic resections is the resemblance of open surgery that it offers to the surgeon [35]. When discussing its benefits during the performance of a Whipple procedure, on the other hand, the additional advantage of surgeon comfort provided by sitting at the console to control the master interface takes precedence. Just as it has been proven in the colorectal literature, the robotic pancreaticoduodenectomy offers the advantages of a decreased length of stay and fewer wound infections or surgical site occurrences, while the oncologic outcomes are comparable to open surgery [36, 37].

Robotic distal pancreatectomy has been studied, too, with excellent results particularly when it comes to splenic preservation due to the dexterity offered by wristed instrumentation and multi-arm control [38]. On the other hand, robotic distal pancreatectomy is equally effective when a splenectomy is performed at the same time [39]. When a comparison is made between the robotic and the laparoscopic approaches, robotic distal pancreatectomy has been shown to have a lower estimated blood loss, a higher spleen preservation rate, and a shorter hospital stay in spite of a longer operative time [40].

Equally demanding and intensive is minimally invasive hepatic surgery. In fact, although experts have shown that the laparoscopic technique is feasible and reproducible in their hands, the robotic platform has allowed them to have greater control of vascular and biliary structures due to its multiple advantages over laparoscopy which have been extensively reviewed. Comparisons between the two methods have been made in the early 2000s with the same conclusions drawn years later [41]. Although wedge resections and segmentectomies have been reported, the most impressive results have been seen with major hepatectomies when their outcomes and metrics have been analyzed in the literature [42, 43]. A subsequent meta-analysis in 2015 comparing robotic and laparoscopic liver resections reported greater blood loss and longer operative time for the robotic approach. However, most likely the blood loss observation had to do with the technique being used at that time. Nevertheless, both techniques were found to be equally efficient in terms of oncologic outcomes, the length of stay, and complication rates [44]. In the same fashion, another review of the literature the same year concluded that robotic hepatic surgery is as effective as laparoscopic and open surgery [45]. A review of the literature specifically dealing with the topic of hepatocellular carcinoma reported a similar statement [46].

No matter how many liver resection series were examined, however, although it may seem disappointing to note that robotic hepatic surgery was not found to be superior to its laparoscopic counterpart, what is essential to realize is that the field is evolving and all of the data support the fact that it is safe, comparable to laparoscopy, and with the same oncologic outcomes in spite of the difficulty level associated with this type of operation.

5. Gastric surgical oncology: refinement takes shape

Another complex type of operation requiring a high skill level is gastric surgery, especially when a neoplastic process is at the core of the situation and the requirement for an extensive lymphadenectomy is essential. Where robotic liver surgery has failed to show superiority on multiple fronts when compared to laparoscopy, gastric surgery has compensated and exceeded the expectations, as seen on an impressive series of 200 consecutive gastric resections published in 2013, including decreased operative time, superior lymph node yield, and decreased length of stay [47]. The Asian literature has extensively published case series such as this with impressive results.

The robotic platform allows the surgeon to overcome some of the limitations presented by laparoscopy, above all when performing a D2 lymphadenectomy, where its multiple advantages become more obvious [48]. The usefulness of the surgical robot has been noticed regarding the performance of robotic-sewn anastomoses and challenging dissections near the gastroesophageal junction and the pyloric region, proving helpful during total gastrectomies, for instance [49]. Overall, the robotic technology has established its relevance in the field of gastric surgical oncology for many reasons and will continue to do so in the near future [50, 51].

A meta-analysis from 2013 has actually established that robotic gastric surgery is superior to its laparoscopic counterpart in terms of estimated blood loss and hospital stay, with the only difference being a longer operative time. However, the benefits have been shown and are more definitive than those seen on liver resection [52]. Another meta-analysis has also supported the validity and superiority of robotic gastrectomy for cancer when compared to open surgery [53]. This subject is so important in the surgical oncology community that a worldwide database was created to track the results from gastric resections corresponding to the robotic, laparoscopic, and open modalities [54].

Another aspect that is interesting to note is the fact that robotic gastric resections may facilitate future laparoscopic resections and decrease the operative time for both approaches once the surgeon's learning curve is mastered. This is in addition to the finding of lower estimated blood loss on the robotic group [55]. In fact, as the learning curve for robotic gastric resections is surpassed, the D2 lymphadenectomy yield improves and is superior to the laparoscopic outcome [56]. These observations made by the experts in this field are a testament to the fact that the robotic technology enables the surgeon to refine the technique to the point that, regardless of the level of difficulty required for this type of procedure, it is possible to continue to improve as the case volume increases. The same conclusion has been drawn from series that include both subtotal and total gastrectomies performed robotically [57].

This refinement of surgical technique is evident when a robotic-sewn anastomosis is created, as mentioned before, which has been found to be reproducible and very convenient in total and subtotal gastrectomies with a Roux en Y, Billroth I, or Billroth II reconstruction, depending on the case [58, 59].

6. Bariatric and antireflux surgery: the youngest field is evolving

In order to discuss the remarkable progress that has been made in the subspecialty of metabolic and bariatric surgery thanks to the robotic technology, it is important to first recognize the surgical robot's applications in antireflux procedures, especially those in which paraesophageal hernia repair is necessary. Such a case is seen with giant paraesophageal hernias, where the complexity of the repair requires a high level of dexterity due to the size of the diaphragmatic defect and the limited space available at the gastroesophageal junction, with vital structures such as the aorta, the inferior vena cava, and the esophagus can be injured, in addition to the spleen and the liver, due to the requirements posed by the tension on the hernia edges. The robotic platform shines in instances such as this, with results that are similar to the laparoscopic rate of complications in expert hands, but with lower hernia recurrence rates [60].

The same observation is true when a redo antireflux operation and hiatal hernia repair are performed robotically. The results are excellent and consistent with the superiority granted by improved dexterity in a field where the normal anatomy has been violated, and where the dissection must resemble what once was expected, structurally speaking [61].

With respect to metabolic and bariatric surgery, robotic surgeons have advanced this continuously evolving field at high speed due to their spirit of innovation and the high level of difficulty caused by their patients' body habitus, which requires them to develop techniques for dissection, exposure, and port placement that would normally not be necessary on patients with a lower body mass index.

A very helpful systematic review has already demonstrated that robotic bariatric surgery is not exclusively favored in redo cases, but is actually being utilized in non-revision operations where a robotic-sewn intracorporeal gastrojejunostomy or jejunojejunostomy anastomosis is constructed during a Roux en Y gastric bypass, or where a challenging gastric resection is necessary during a sleeve gastrectomy. In fact, even if the surgeons choose to staple the anastomoses during Roux en Y gastric bypass, the robotic technology enables them to perform the enterotomy or gastrotomy closure more efficiently [62].

Of course, the relevance of the robotic approach has been exposed in the unusually complex arena of bariatric surgery revisions, where the experts have been able to achieve results with more advanced dexterity and with a more ergonomically feasible method, with excellent visibility and with the advantage offered by the ability to control three arms and the endoscope simultaneously [63, 64].

It is important to note that robotic bariatric surgery has also been found to be relevant in the super obese patients who undergo a sleeve gastrectomy. In these complex cases, with BMI > 50, the robotic technology has proven very useful for the multiple reasons that have been exposed above for bariatric revision operations. This is interesting to realize, since typically most non-bariatric surgeons associate the surgical robot with the Roux en Y gastric bypass and revision surgery. In fact, the robotic approach may increase the surgeon's skill level to then undertake a difficult gastric bypass or a revision procedure while building on the experience offered by robotic sleeve gastrectomy [65].

As expected, when the most technically demanding bariatric operations are performed, the robotic approach takes precedence, as demonstrated by the creation of intracorporeal anastomoses during revision cases where a conversion from a failed sleeve gastrectomy to a duodenoileal bypass is carried out, both in a classic duodenal switch, as well as in a single-anastomosis duodenal switch (SADI), to give an example [66].

7. Pediatric surgery: applications in spite of size

It is remarkable to realize that the robotic platform has been successfully applied to the pediatric population, where the limitations imposed by size have been partially overcome by the robotic system's well-established advantages over conventional laparoscopy.

While the purpose of this chapter is not to discuss robotic surgery applications in the pediatric population in depth, the goal of this section is to document some of the work that has been done in the subspecialty of Pediatric Surgery with the robotic technology, especially with the da Vinci system.

An important pediatric surgery review that was presented at an international conference in 2007, and published in 2008, showed how the most common robotic surgery applications include but are not limited to pyeloplasty, fundoplication, and patent ductus arteriosus ligation. The authors concluded that although the operative time was longer when compared to laparoscopy, they preferred the robotic platform for the same reasons that their non-pediatric surgeon colleagues have described over years. On the other hand, they expressed their concern regarding the need to make this equipment suitable for neonates and to decrease the cost associated with these operations when the technology is used [67].

A more specialized use of the robotic system in pediatric surgery has been described with excellent results comparable to the open approach for choledochal cyst excision and biliary reconstruction [68]. This is just an example of what can be achieved by members of the surgical community who continue to innovate in their fields when they remain open to progress in a responsible manner.

8. Endocrine surgery: robotics in unusual places

Even the subspecialty of Endocrine Surgery has witnessed the advancement of robotics, both in the retroperitoneum with adrenalectomy and in the neck with thyroidectomy. This is a controversial area, especially regarding endocrine surgery of the neck with the surgical robot, yet some experts continue to perform their operations safely. A major criticism for the use of the robot in the neck is the fact that it requires such a high level of skill that it should only be reserved to the experts.

However, with respect to adrenalectomy, the robotic system can be used via the posterior retroperitoneal approach and the lateral transperitoneal approach. The latter is favored for larger tumors. In fact, some authors favor the lateral transperitoneal approach for most tumors regardless of their size and prefer to apply it to pregnant women and children [69].

When compared to laparoscopy, robotic adrenalectomy has been determined to be as effective and to have the same rate of complications, but its major disadvantage is the cost associated with the procedure when the robotic platform is used. Nevertheless, it is a safe technique and the conversion rate to open surgery is very low [70]. In fact, a more recent literature review has shown that robotic adrenalectomy, when performed at high-volume centers, has superior results to the laparoscopic approach, with lower estimated blood loss, shorter hospital stay, and improvement in intra-operative time with a higher case volume [71]. This is an improvement over a prior meta-analysis published 2 years earlier which had concluded that there is no advantage of robotic adrenalectomy over laparoscopic adrenalectomy [72].

On the topic of thyroidectomy, the robotic technique has been found to be very advantageous to the surgeons due to superior ergonomics when compared to the endoscopic approach, in addition to the fact that the learning curve is easier to master with the surgical robot [73]. Another concept was introduced by a group that reported on their initial experience with robotic thyroidectomy in 2011, which was the fact that the robotic technique eliminates the need to have an assistant in spite of an increased procedure time [74].

A recent literature review dedicated to the study of prior series of robotic thyroidectomy for cancer and their comparison to the open approach concluded that the open technique is superior in terms of oncologic outcomes, decreased operative times, and lower cost. However, the robotic approach was comparable to open thyroidectomy for cancer regarding morbidity, short-term recurrence rates, and quality of life outcomes. The authors warn that the technique for this indication should be reserved to the experts at high-volume centers [75]. A few years earlier, a large case series of robotic thyroidectomy for cancer had precisely shown that the robotic approach has decreased operative times and improved lymph node yields compared to the endoscopic technique. Moreover, the robotic learning curve was shorter [76]. A large case series of 100 patients with papillary thyroid microcarcinoma was published and reported on the robotic total thyroidectomy with central node dissection while compared to the open approach. The results were comparable to the open group, with no conversions, and with similar lymph node retrieval [77]. This is just an example of how far some experts have contributed to the advancement of this subspecialty with a minimally invasive technique that has surpassed its endoscopic counterpart.

9. Cholecystectomy: from the traditional to single-site

Robotic cholecystectomy is often one of the first procedures that surgeons learn to perform with the robot in order to overcome their learning curve and build a basic skill set that will allow them to embark on challenging operations in the future [78]. However, it is also true that some cholecystectomies may become complex operations that may lead to complications when meticulous technique and sound surgical judgment are not applied. The initial years of robotics in General Surgery were times when some groups advocated for performing this procedure only for training purposes since there appeared to be no value over the well-established laparoscopic technique, which had been the gold standard for a long time.

A year later, another group presented their data on robotic cholecystectomy by using a different port arrangement in the lower abdominal wall, separate from the traditional approach in laparoscopic cholecystectomy. The results were satisfactory, with safety and efficiency being at the core of their manuscript [79]. Subsequent case series by different centers published the data corresponding to the first robotic single-site cholecystectomies performed at their institution. The common conclusion was, as expected, that this technique was safe and feasible, and that the learning curve is relatively easy to overcome. On the other hand, surgical resident training did not affect the results in a negative way [80, 81].

The technique consists of using a single-site port with four channels created by intuitive Surgical to overcome the limitation that arises from laparoendoscopic single-site (LESS) surgery when the arm movement is the opposite of what the surgeon expects due to the need to pivot the instruments around a central axis. With the robotic single-site technology, however, although there is no wristed articulation of the instruments, the limitation is overcome when the surgeon sits at the console and the arm movement is inverted so that the instrument movement matches the hand movement at the console. This can be very convenient and, indeed, can be applied to perform single-site cholecystectomy in patients with a high BMI most of the time (see **Figures 4–7**).

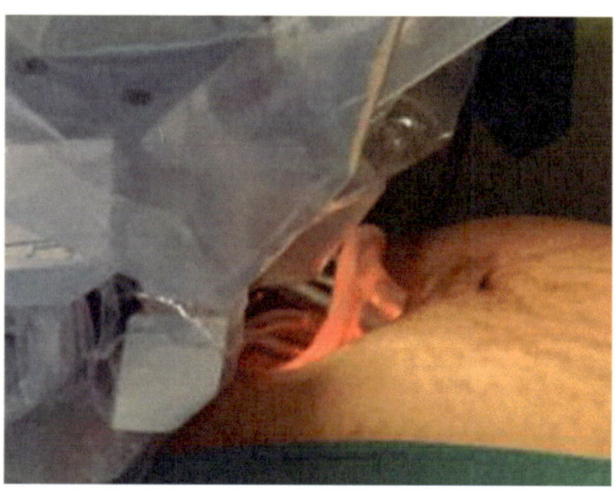

Figure 4. da Vinci single-site port inserted through an infraumbilical 2.5 cm incision.

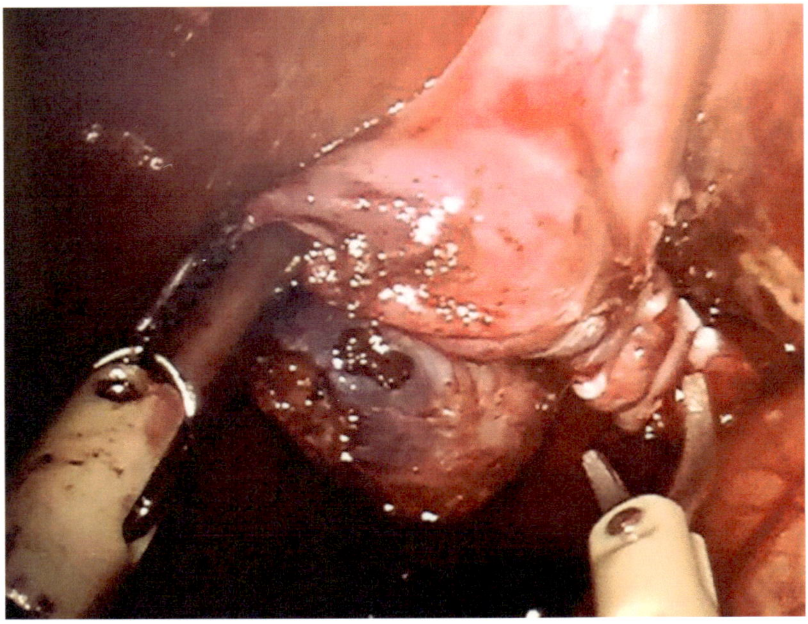

Figure 5. da Vinci single-site instruments in action during a robotic cholecystectomy.

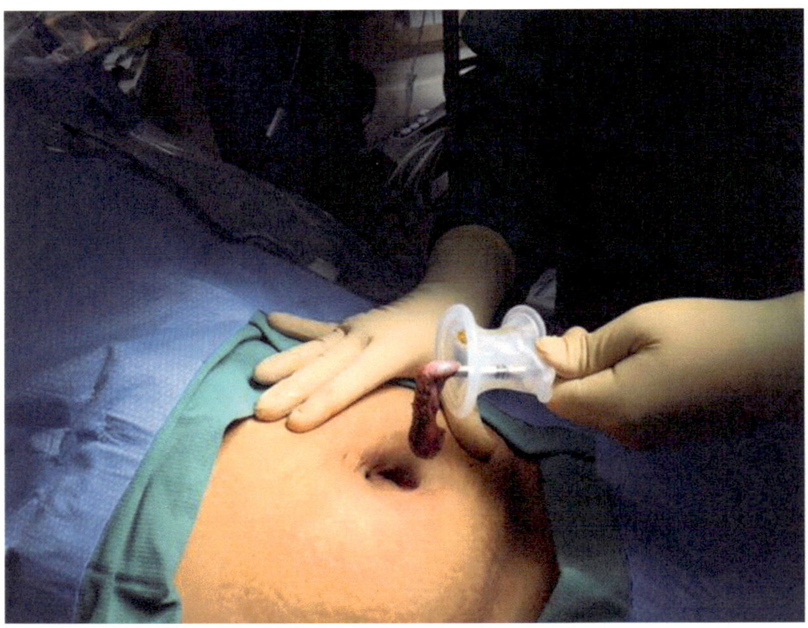

Figure 6. Specimen extracted via the only incision.

http://dx.doi.org/10.5772/intechopen.68864

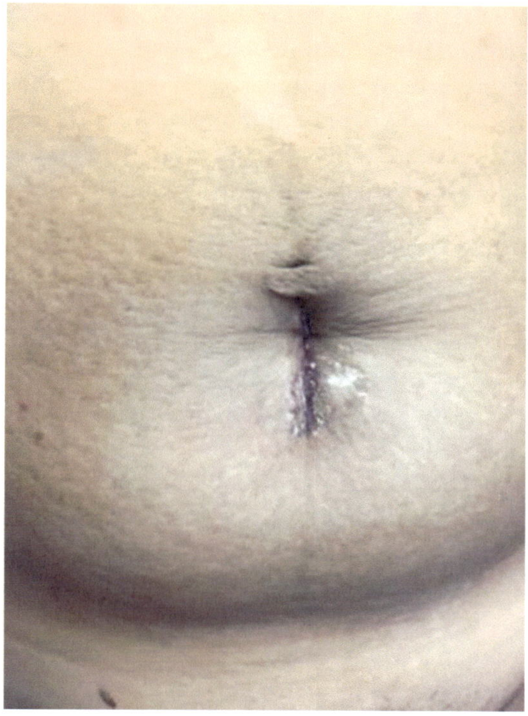

Figure 7. Final cosmetic result. Based on the patient's abdominal wall thickness and BMI, sometimes a vertical skin incision is necessary, although a transverse skin incision is made in most cases. A vertical fascial incision is always favored.

10. Hernia repair: from closing defects to suturing mesh

The field of hernia repair and abdominal wall reconstruction has seen an increasing amount of studies and case series recently published which present new techniques that continue to advance the subspecialty of minimally invasive abdominal wall reconstruction. The results are outstanding and the surgeons witness them to the point that patient satisfaction correlates with less chronic pain and decreased hospital stay. Although laparoscopic hernia repair has been established as an appropriate technique in most cases, its Achilles heel has always been the presence of chronic pain, most likely due to transfascial sutures and to the utilization of tacks for intraperitoneal mesh fixation, whether they are permanent or absorbable.

On the subject of intraperitoneal mesh fixation, a study published in 2012 presented excellent results when the primary ventral hernia defect was closed with intracorporeal sutures placed with the robotic system, and when the mesh was fixed as an underlay with circumferential sutures, without the use of tacks [82]. Just to compare, as early as 2003 another manuscript had already presented a robotic hernia repair, but the idea at that time was to still secure the mesh with tacks and to not close the primary defect with sutures [83]. As it can be seen, therefore, the field of hernia repair has come a long way by establishing the new concept of primary defect closure for the sake of a more mechanically and physiologically normal abdominal wall, and avoidance of transfascial sutures and tacks to prevent chronic pain. Furthermore, all of the series have determined that the robotic platform offers the opportunity to perform enterolysis more efficiently through the multiple benefits that have been described before [84–86] (see **Figures 8–10**).

Regarding the specific situation of inguinal hernia repair, which has been extensively performed with the laparoscopic total extraperitoneal (TEP) and the transabdominal preperitoneal (TAPP) approaches, the robotic technique offers remarkable advantages in the confined space where it takes place, including the dexterity offered by the wristed instruments and the ability to perform a finer dissection and suture the peritoneal flap in the case of a TAPP. The Urology literature recognizes the relevance of the surgical robot when a TEP is performed at the time of robotic prostatectomy as a combined operation [87]. In the General Surgery literature, where the robotic TAPP approach is favored, the absence of neuralgia after the operation is likely a reflection of all of the advantages offered by the robotic platform in addition to the avoidance of tacks to fix the mesh and close the peritoneal flap, which is similar to the observation made in the ventral hernia series when tacks are avoided as well as transfascial sutures [88]. In addition, the robotic technology has been used to develop new minimally invasive ways to reconstruct the abdominal wall, such as the robotic transversus abdominis release as a posterior component separation with the preperitoneal placement of mesh, but the description of all of these techniques is beyond the scope of this chapter. In reality, such monumental task deserves a separate chapter in a future publication.

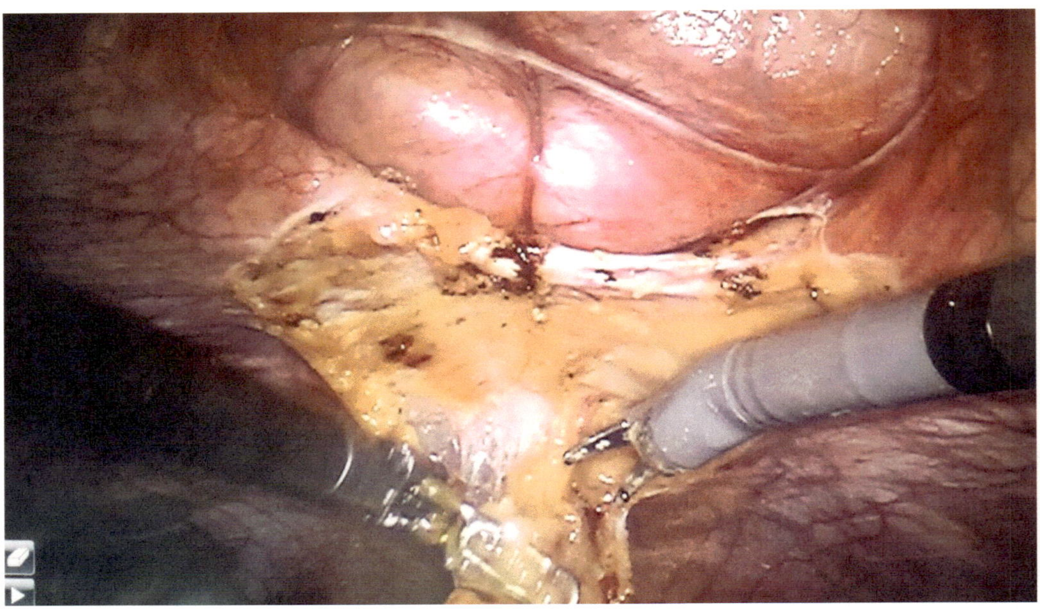

Figure 8. Robotic enterolysis in anticipation of primary closure of an incisional ventral hernia defect, and prior to intraperitoneal mesh implantation with circumferential intracorporeal sutures.

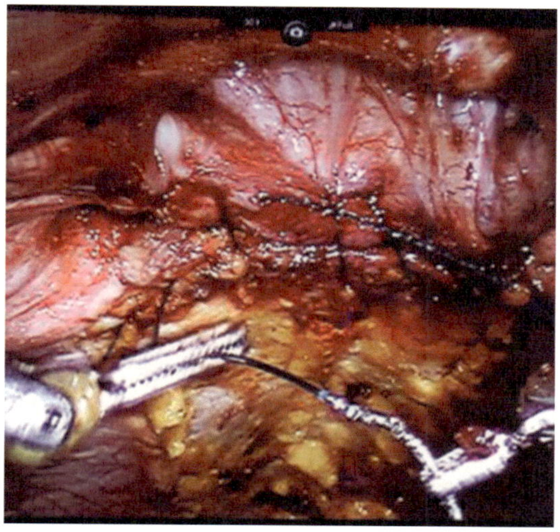

Figure 9. Intracorporeal robotic suturing for closure of incisional ventral hernia defect.

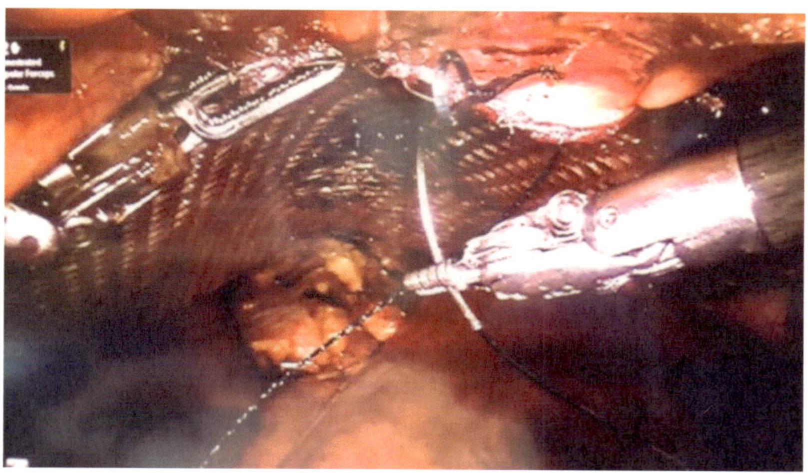

Figure 10. Circumferential intracorporeal suturing of mesh for fixation while avoiding the use of tacks or transfascial sutures.

11. General surgery: robotics applied to all cases

While it is true that much of the progress made in robotic surgery has originated from multiple case series in the surgical subspecialties, as it has been extensively documented in this chapter, a significant degree of advancement has come from true General Surgery programs that have continued to perfect the technique and its applications in a vast range of procedures with success [89]. The perfect example came from an extensive case series of robotic General Surgery cases in a large European community hospital. What is significant about this publication is the fact that it was 2003 and, above all, the relevant observation that the 207 procedures were performed with the surgical technology in the community hospital environment or, in other words, not in an academic institution associated with a university. Of course, being a large hospital, it was a referral center for other hospitals in the region, but it was a community institution after all [90]. Another European case series of 94 patients was published in 2007 with similar results and conclusions [91].

These studies served as an inspiration for other surgeons who wished to incorporate the robotic surgical approach to their armamentarium and to offer the benefits of robotic surgery to their patients in the General Surgery environment, with most of the series favoring gastrointestinal surgery [92, 93]. Perhaps one of the first publications to lay the foundation for the need to include hospital administrators, medical school and residency program authorities, and the surgical team leadership in the process of creating a successful Robotic General Surgery robotic program was an American manuscript from 2010 [94]. This manuscript opened the gate to a new level of discussion that needed to begin in order to establish the guidelines for a successful, productive, safe, and efficient robotic program to thrive.

12. The last argument: innovation cannot be stopped

In 2016, a comprehensive review of all surgical specialties (such as Urology, Gynecology, and Thoracic Surgery) and General Surgery subspecialties (presented in this chapter) included cases performed from 2000 to 2013. Adverse events were analyzed, and the conclusion was that they were less frequent in those specialties where the surgical robot is used more often. Most of the events were due to equipment malfunction, however, and not to surgeon technique [95]. Nonetheless, once again, surgical judgment takes priority and should always be the driving force in control of the surgical robot.

As long as the advanced technology is utilized to impact our patients in a positive way, there will always be the risk of complications, and no surgeon can deny that, whether the approach is open, laparoscopic, or robotic. In fact, in 2013, a European study expanded on the topic of guidelines and principles that are necessary to guide a successful robotic surgery program. The elements for the ideal organizational model to implement such an efficient program were discussed, but what seems to be different from prior publications by other groups is the fact that the investigators suggested the expansion of the robotic platform to more subspecialties in General Surgery [96]. This is a shifting paradigm from the old idea that the surgical robot should only be reserved to perform highly specialized procedures such as colorectal, complex foregut, or hepatobiliary, pancreatic, and gastric oncology.

Innovation cannot be stopped. When surgeons keep their patients' safety in mind as their top priority, safe innovation becomes a reflection of progress in their specialty. Human beings have always been creative, and their creativity will continue to be applied in their profession regardless of opposition from those who prefer the status quo because it is more comfortable to do so.

The first American case series of robotic General Surgery cases in a community hospital to this date, to this author's knowledge, did not come from a tertiary referral center or fully academic institution. It was inspired by prior European series from the early and mid-2000s that have already been presented in this chapter. The first American case series, however, came from a very small acute care community hospital of 266 beds affiliated with a university, but lacking a residency program and consisting of a single surgeon experience. The total number of procedures performed was 101, with case #101 being meaningful to the surgeon and his team because of its relevance as the first robotic bariatric operation performed in the city [97].

This publication from 2016 has paved the way for future case series where a higher volume of cases is necessary to achieve statistical significance and inspire others to conduct randomized controlled trials in the future. In fact, a follow-up study is already being prepared for the first 200 robotic General Surgery cases in the same community hospital, this time with statistical significance due to the larger size of the series.

While multiple case series have been reported in the United States, none has included a large variety of cases across most surgical subspecialties including hernia, colorectal, gallbladder, foregut, and bariatric surgery, particularly in a community hospital environment where resources are limited and with the da Vinci S system being used to perform these operations from 2014 to 2015. The manuscript's most important conclusion is twofold: first and foremost, a successful robotic General Surgery program can be implemented in a community hospital by training the surgical team as the surgeon overcomes the learning curve, with improved results seen as the number of cases increases. Secondly, and what may be the most important observation, the study suggests that the surgical robot can be safely and efficiently used both for complex and simple General Surgery procedures, not just for the complex cases.

In conclusion, while hoping to stimulate the international surgical community to appreciate the value of the surgical robot for General Surgery and its multiple subspecialties, this author's ultimate goal is to remind himself and his colleagues around the world that the only way to improve is to continue to learn, both from our own mistakes as well as from the substantial body of knowledge that has been compiled over the years. This is the legacy left for us by a few pioneers who began to open their minds and think outside the dogma that had been established as the infallible truth: that laparoscopy is the least invasive way to perform an operation, and that nothing else can be created that will improve upon its benefits. Innovation, at the core of every surgeon's mind and spirit, will continue to advance in our field to benefit our patients. The best decision we can make today is to prepare ourselves to join others in this magnificent enterprise without being left behind. After all, our patients deserve our best effort to improve and to learn until our last breath.

Author details

Rodolfo José Oviedo Barrera

Address all correspondence to: rodolfo.j.oviedo@gmail.com

General, Minimally Invasive, Robotic, Oncologic, Metabolic and Bariatric Surgery, Florida State University College of Medicine, Tallahassee, Florida, USA

References

[1] Hockstein NG, Gourin CG, Faust RA, Terris DJ. A history of robots: From science fiction to surgical robotics. Journal of Robotic Surgery. 2007;**1**:113-118

[2] Shah J, Vyas A, Vyas D. The history of robotics in surgical specialties. American Journal of Robotic Surgery. 2014;**1**(1):12-20

[3] Intuitive Surgical. History. https://www.intuitivesurgical.com/company/history/. Accessed February 25, 2017

[4] Antoniou SA, Antoniou GA, Antoniou AI, Granderath FA. Past, present, and future of minimally invasive abdominal surgery. JSLS. 2015;**19**(3):e2015.00052

[5] Lanfranco AR, Castellanos AE, Desai JP, Meyers WC. Robotic surgery: A current perspective. Annals of Surgery. 2004;**239**:14-21

[6] Avgousti S, et al. Medical telerobotic systems: Current status and future trends. BioMedEng Online. 2016;**15**(96):1-44. DOI: 1186/s12938-016-0217-7

[7] van der Meijden OAJ, Schijven MP. The value of haptic feedback in conventional and robot-assisted minimal invasive surgery and virtual reality training: A current review. Surgical Endoscopy. 2009;**23**:1180-1190

[8] Lendvay TS, et al. Virtual reality robotic surgery warm-up improves task performance in a dry lab environment: A prospective randomized controlled study. Journal of the American College of Surgeons. 2013;**216**(6):1181-1192

[9] Chitwood WR, et al. Robotic surgical training in an academic institution. Journal of the American College of Surgeons. 2001;**234**(4):475-486

[10] Guerrieri M, Campagnacci R, Sperti P, Belfiori G, Gesuita R, Ghiselli R. Totally robotic vs 3D laparoscopic colectomy: A single center's preliminary experience. World Journal of Gastroenterology. 2015;**21**(46):13152-13159

[11] Pucci MJ, Beekley AC. Use of robotics in colon and rectal surgery. Clinics in Colon and Rectal Surgery. 2013;**26**:39-46

[12] Weaver A, Steele S. Robotics in colorectal surgery. F1000Research. 2016;**5**(F1000 Faculty Rev):2373. DOI: 10.12688/f1000research.9389.1

[13] Trastulli S, et al. Robotic versus laparoscopic approach in colonic resections for cancer and benign diseases: Systematic review and meta-analysis. PLoS ONE. 2015;**10**(7):e0134062. DOI: 10.1371/journal.pone.0134062

[14] Lim DR, et al. Robotic versus laparoscopic anterior resection of sigmoid colon cancer: Comparative study of long term oncologic outcomes. Surgical Endoscopy. 2013;**27**:1379-1385

[15] de'Angelis N, Alghamdi S, Renda A, Azoulay D, Brunetti F. Initial experience of robotic versus laparoscopic colectomy for transverse colon cancer: A matched case-control study. World Journal of Surgical Oncology. 2015;**13**:295. DOI: 10.1186/s12957-015-0708-1

[16] Liao G, et al. Robotic-assisted versus laparoscopic colorectal surgery: A meta-analysis of four randomized control trials. World Journal of Surgical Oncology. 2014;**12**:122. http://www.wjso.com/content/12/1/122

[17] Xu H, et al. Robotic versus laparoscopic right colectomy: A meta-analysis. World Journal of Surgical Oncology. 2014;**12**:274. www.wjso.com/content/12/1/274

[18] Trinh BB, Hauch HT, Buell JF, Kandil E. Robot-assisted versus standard laparoscopic colorectal surgery. JSLS. 2014;**18**(4):e2014.00154. DOI: 10.4293/JSLS.2014.00154

[19] Davis BR, Yoo AC, Moore M, Gunnarsson C. Robotic-assisted versus laparoscopic colectomy: Cost and clinical outcomes. JSLS. 2014;**18**:211-224

[20] Liao G, Li YB, Zhao Z, Li X, Deng H, Li G. Robotic-assisted surgery versus open surgery in the treatment of rectal cancer: The current evidence. Scientific Reports. 2016;**6**:26981. DOI: 10.1038/srep26981

[21] Bedirli A, Salman B, Yuksel O. Robotic versus laparoscopic resection for mid and low rectal cancers. JSLS. 2016;**20**(1):e2015.00110. DOI: 10.4293/JSLS.2015.00110

[22] Wing Chung Mak T, Fung Yee Lee J, Futaba K, SokFei Hon S, Kwok Yu Ngo D, Siu Man Ng S. Robotic surgery for rectal cancer: A systematic review of current practice. World Journal of Gastrointestinal Oncology. 2014;**6**(6):184-193

[23] Bae SU, Min BS, Kim NK. Robotic low ligation of the inferior mesenteric artery for rectal cancer using the Firefly technique. Yonsei Medical Journal. 2015;**56**(4):1028-1035

[24] Patriti A, Ceccarelli G, Bartoli A, Spaziani A, Biancafarina A, Casciola L. Short- and medium-term outcome of robot-assisted and traditional laparoscopic rectal resection. JSLS. 2009;**13**:176-183

[25] Bianchi PP, Petz W, Luca F, Biffi R, Spinoglio G, Montorsi M. Laparoscopic and robotic total mesorectal excision in the treatment of rectal cancer. Brief review and personal remarks. Frontiers in Oncology. 2014;**4**(98). DOI: 10.3389/fonc.2014.00098

[26] Park EJ, et al. Is the learning curve of robotic low anterior resection shorter than laparoscopic low anterior resection for rectal cancer? A comparative analysis of clinicopathologic outcomes between robotic and laparoscopic surgeries. Medicine. 2014;**93**(25):e109. DOI: 10.1097/MD.0000000000000109

[27] Bae SU, et al. Robotic and laparoscopic pelvic lymph node dissection for rectal cancer: Short-term outcomes of 21 consecutive series. Annals of Surgical Treatment and Research. 2014;**86**(2):76-82

[28] Byrn JC, Hrabe JE, Charlton ME. An initial experience with 85 consecutive robotic assisted rectal dissections: Improved operating times and lower costs with experience. Surgical Endoscopy. 2014;**28**(11):3101-3107

[29] Maciel V, et al. Diverticular disease complicated with colovesical fistula: laparoscopic versus robotic management. International Surgery. 2014;**99**:203-210

[30] deHoog DENM, Heemskerk J, Nieman FHM, van Gemert WG, Baeten CGMI, Bouvy ND. Recurrence and functional results after open versus conventional laparoscopic versus robot-assisted laparoscopic rectopexy for rectal prolapse: A case-control study. International Journal of Colorectal Disease. 2009;**24**:1201-1206

[31] Heemskerk J, de Hoog DENM, van Gemert WG, Baeten CGMI, Greve JWM, Bouvy ND. Robot-assisted vs. conventional laparoscopic rectopexy for rectal prolapse: A comparative study on costs and time. Diseases of the Colon & Rectum. 2007;**50**:1825-1830

[32] Zureikat AH, Moser AJ, Boone BA, Bartlett DL, Zenati M, Zeh HJ. 250 Robotic pancreatic resections: Safety and feasibility. Annals of Surgery. 2013;**258**(4):554-562

[33] Wayne M, Steele J, Iskandar M, Cooperman A. Robotic pancreatic surgery is no substitute for experience and clinical judgment: An initial experience and literature review. World Journal of Surgical Oncology. 2013;**11**:160. http://www.wjso.com/content/11/1/160

[34] Joyce D, Morris-Stiff G, Falk GA, El-Hayek K, Chalikonda S, Walsh RM. Robotic surgery of the pancreas. World Journal of Gastroenterology. 2014;**20**(40):14726-14732

[35] Del Chiaro M, Segersvard R. The state of the art of robotic pancreatectomy. BioMed Research International. 2014;**920492**:1-5. http://dx.doi.org/10.1155/2014/920492

[36] Baker EH, et al. Robotic pancreaticoduodenectomy for pancreatic adenocarcinoma: Role in 2014 and beyond. Journal of Gastrointestinal Oncology. 2015;**6**(4):396-405

[37] Zhan Q, et al. Outcomes of robotic surgery for pancreatic ductal adenocarcinoma. Chinese Journal of Cancer. 2015;**27**(6):604-610

[38] Liu Y, et al. Robotic spleen-preserving laparoscopic distal pancreatectomy: A single-centered Chinese experience. World Journal of Surgical Oncology. 2015;**13**:275. DOI: 10.1186/s12957-015-0671-x

[39] Ryan CE, Ross SB, Sukharamwala PB, Sadowitz BD, Wood TW, Rosemurgy AS. Distal pancreatectomy and splenectomy: A robotic or LESS approach. JSLS. 2015;**19**(1):e2014.00246. DOI: 10.4293/JSLS.2014.00246

[40] Zhou JY, et al. Robotic versus laparoscopic distal pancreatectomy: A meta-analysis of short-term outcomes. PLoS ONE. 2016;**11**(3):e0151189. DOI: 10.1371/journal.pone.0151189

[41] Choi SB, et al. Early experiences of robotic-assisted laparoscopic liver resection. Yonsei Medical Journal. 2008;**49**(4):632-638

[42] Reggiani P, Antonelli B, Rossi G. Robotic surgery of the liver: Italian experience and review of the literature. ecancer. 2013;**7**:358. DOi: 10.3332/ecancer.2013.358

[43] Ho CM, Wakabayashi G, Nitta H, Ito N, Hasegawa Y, Takahara T. Systematic review of robotic liver resection. Surgical Endoscopy. 2013;**27**:732-739

[44] Montalti R, Berardi G, Patriti A, Vivarelli M, Troisi RI. Outcomes of robotic vs laparoscopic hepatectomy: A systematic review and meta-analysis. World Journal of Gastroenterology. 2015;**21**(27):8441-8451

[45] Giulianotti PC, Bianco FM, Daskalaki D, Gonzalez-Ciccarelli LF, Kim J, Benedetti E. Robotic liver surgery: Technical aspects and review of the literature. Hepatobiliary Surgery and Nutrition. 2016;**5**(4):311-321

[46] Levi Sandri GB, et al. Laparoscopic and robotic approach for hepatocellular carcinoma—state of the art. Hepatobiliary Surgery and Nutrition. 2016;**5**(6):478-484

[47] Park JY, Kim YW, Ryu KW, Eom BW, Yoon HM, Reim D. Emerging role of robot-assisted gastrectomy: Analysis of consecutive 200 cases. Journal of Gastric Cancer. 2013;**13**(4):255-262

[48] Caruso S, et al. Laparoscopic and robot-assisted gastrectomy for gastric cancer: Current considerations. World Journal of Gastroenterology. 2016;**22**(25):5694-5717

[49] Baek SJ, Lee DW, Park SS, Kim SH. Current status of robot-assisted gastric surgery. World Journal of Gastrointestinal Oncology. 2011;**3**(10):137-143

[50] Caruso S, et al. Robot-assisted laparoscopic gastrectomy for gastric cancer. World Journal of Gastrointestinal Endoscopy. 2017;**9**(1):1-11

[51] Procopiuc L, Tudor S, Manuc M, Diculescu M, Vasilescu C. Robot-assisted surgery for gastric cancer. World Journal of Gastrointestinal Oncology. 2016;**8**(1):8-17

[52] Marano A, Choi YY, Hyung WJ, Kim YM, Kim J, Noh SH. Robotic versus laparoscopic versus open gastrectomy: A meta-analysis. Journal of Gastric Cancer. 2013;**13**(3):136-148

[53] Liao G, et al. Robotic versus open gastrectomy for gastric cancer: A meta-analysis. PLoS ONE. 2013;**8**(12):e81946. DOI: 10.1371/journal.pone.0081946

[54] Desiderio J, et al. Robotic, laparoscopic and open surgery for gastric cancer compared on surgical, clinical and oncological outcomes: A multi-institutional chart review. A study protocol of the International study group on Minimally Invasive surgery for GASTRIc Cancer—IMIGASTRIC. BMJ Open. 2015;**5**:e008198. DOI: 10.1136/bmjopen-2015-008198

[55] Huang KH, et al. Comparison of the operative outcomes and learning curves between laparoscopic and robotic gastrectomy for gastric cancer. PLoS ONE. 2014;**9**(10):e111499. DOI: 10.1371/journal.pone.0111499

[56] Cianchi F, et al. Robotic vs laparoscopic distal gastrectomy with D2 lymphadenectomy for gastric cancer: A retrospective comparative mono-institutional study. BMC Surgery. 2016;**16**:65. DOI: 10.1186/s12893-016-0180-z

[57] Zong L, Seto Y, Aikou S, Takahashi T. Efficacy evaluation of total and subtotal gastrectomies in robotic surgery for gastric cancer compared with that in open and laparoscopic resections: A meta-analysis. PLoS ONE. 2014;**9**(7):e103312. DOI: 10.1371/journal.pone.0103312

[58] Liu XX, Jiang ZW, Zhao Y, Pan HF, Li JS. Full robot-assisted gastrectomy with intracorporeal robot-sewn anastomosis produces satisfying outcomes. World Journal of Gastroenterology. 2013;**19**(38):6427-6437

[59] Parisi A, et al. Robotic total gastrectomy with intracorporeal robot-sewn anastomosis. Medicine. 2015;**94**(49):e1922. DOI: 10.1097/MD.0000000000001922

[60] Seetharamaiah R, et al. Robotic repair of giant paraesophageal hernias. JSLS. 2013;**17**:570-577

[61] Tolboom RC, Draaisma WA, Broeders IAMJ. Evaluation of conventional laparoscopic versus robot-assisted laparoscopic redo hiatal hernia and antireflux surgery: A cohort study. Journal of Robotic Surgery. 2016;**10**:33-39

[62] Cirocchi R, et al. Current status of robotic bariatric surgery: A systematic review. BMC Surgery. 2013;**13**:53. http://www.biomedcentral.com/1471-2482/13/53

[63] Bindal V, et al. Review of contemporary role of robotics in bariatric surgery. Journal of Minimal Access Surgery. 2015;**11**(1):16-21

[64] Acquafresca PA, Palermo M, Rogula T, Duza GE, Serra E. Most common robotic bariatric procedures: Review and technical aspects. Journal of Minimal Access Surgery. 2015;**9**:9. DOI: 10.1186/s13022-015-0019-9

[65] Bhatia P, et al. Robot-assisted sleeve gastrectomy in morbidly obese versus super obese patients. JSLS. 2014;**18**(3):e2014.00099. DOI: 10.4293/JSLS.2014.00099

[66] Vilallonga R, et al. Robotically assisted single anastomosis duodenoileal bypass after previous sleeve gastrectomy implementing high valuable technology for complex procedures. Journal of Obesity. 2015;**586419**. http://dx.doi.org/10.1155/2015/586419

[67] Sinha CK, Haddad M. Robot-assisted surgery in children: Current status. Journal of Robotic Surgery. 2008;**1**:243-246

[68] Kim NY, et al. Retrospective assessment of the validity of robotic surgery in comparison to open surgery for pediatric choledochal cyst. Yonsei Medical Journal. 2015;**56**(3):737-743

[69] Okoh AK, Berber E. Laparoscopic and robotic adrenal surgery: Transperitoneal approach. Gland Surgery. 2015;**4**(5):435-441

[70] Akarsu C, et al. The early results of our initial experience with robotic adrenalectomy. Turkish Journal of Surgery. 2014;**30**:28-33

[71] Teo XL, Lim SK. Robotic assisted adrenalectomy: Is it ready for prime time? Investigative and Clinical Urology. 2016;**57**(2):S130-146

[72] Chai YJ, et al. Systematic review of surgical approaches for adrenal tumors: Lateral transperitoneal versus posterior retroperitoneal and laparoscopic versus robotic adrenalectomy. International Journal of Endocrinology. 2014;**2014**:918346. http://dx.doi.org/10.1155/2014/918346

[73] Lee J, Chung WY. Robotic surgery for thyroid disease. European Thyroid Journal. 2013;**2**:93-101

[74] Lang BHH, Chow MP. A comparison of surgical outcomes between endoscopic and robotically assisted thyroidectomy: The authors' initial experience. Surgical Endoscopy. 2011:**25**:1617-1623

[75] Liu SYW, Ng EKW. Robotic versus open thyroidectomy for differentiated thyroid cancer: An evidence-based review. International Journal of Endocrinology. 2016;**2016**:4309087. http://dx.doi.org/10.1155/2016/4309087

[76] Lee J, Yun JH, Choi UJ, Kang SW, Jeong JJ, Chung WY. Robotic versus endoscopic thyroidectomy for thyroid cancers: A multi-institutional analysis of early postoperative outcomes and surgical learning curves. Journal of Oncology. 2012;**2012**:734541. DOI: 10.1155/2012/734541

[77] He QQ, et al. Comparative study between robotic total thyroidectomy with central lymph node dissection via bilateral axillo-breast approach and conventional open procedure for papillary thyroid microcarcinoma. Chinese Medical Journal. 2016;**129**(18):2160-2166

[78] Jayaraman S, Davies W, Schlachta CM. Getting started with robotics in general surgery with cholecystectomy: The Canadian experience. Canadian Journal of Surgery. 2009;**52**(5):374-378

[79] Kim JH, et al. Robotic cholecystectomy with new port sites. World Journal of Gastroenterology. 2013;**19**(20):3077-3082

[80] Ayloo S, Choudhury N. Single-site robotic cholecystectomy. JSLS. 2014;**18**(3):e2014.00266. DOI: 10.4293/JSLS.2014.00266

[81] Chung PJ, et al. Single-site robotic cholecystectomy at an inner-city academic center. JSLS. 2015;**19**(3):e2015.00033. DOI: 10.4293/JSLS.2015.00033

[82] Allison N, Tieu K, Snyder B, Pigazzi A, Wilson E. Technical feasibility of robot-assisted ventral hernia repair. World Journal of Surgery. 2012;**36**(2):447-452

[83] Ballantyne GH, Hourmont K, Wasielewski A. Telerobotic laparoscopic repair of incisional ventral hernias using intraperitoneal prosthetic mesh. JSLS. 2003;**7**:7-14

[84] Kudsi OY, Paluvoi N, Bhurtel P, McCabe Z, El-Jabri R. Robotic repair of ventral hernias: Preliminary findings of a case series of 106 consecutive cases. American Journal of Robotic Surgery. 2015;**2**(1):22-26

[85] Gonzalez AM, Romero RJ, Seetharamaiah R, Gallas M, Lamoureux J, Rabaza JR. Laparoscopic ventral hernia repair with primary closure versus no primary closure: Potential benefits of the robotic technology. The International Journal of Medical Robotics and Computer Assisted Surgery. 2015;**11**(2):120-125

[86] Tayar C, Karoui M, Cherqui D, Fagniez PL. Robot-assisted laparoscopic mesh repair of incisional hernias with exclusive intracorporeal suturing: A pilot study. Surgical Endoscopy. 2007;**21**(10):1786-1789

[87] Qazi HAR, et al. Robot-assisted laparoscopic total extraperitoneal hernia repair during prostatectomy: Technique and initial experience. Central European Journal of Urology. 2015;**68**:240-244

[88] Engan C, Engan M, Bonilla V, Dyer DC, Randall BR. Description of robotically assisted single-site transabdominal preperitoneal (RASS-TAPP) inguinal hernia repair and presentation of clinical outcomes. Hernia. 2015;**19**(3):423-428

[89] Dhumane PW, Diana M, Leroy J, Marescaux J. Minimally invasive single-site surgery for the digestive system: A technological review. Journal of Minimal Access Surgery. 2011;**7**(1):40-51

[90] Giulianotti PC, et al. Robotics in general surgery. Personal experience in a large community hospital. JAMA Surgery. 2003;**138**:777-784

[91] Soravia C, et al. Laparoscopic robotic-assisted gastrointestinal surgery: The Geneva experience. Journal of Robotic Surgery. 2008;**1**:291-295

[92] Patriti A, Addeo P, Buchs N, Casciola L, Morel P. Advanced applications of robotics in digestive surgery. Translational Medicine @ UniSa. 2011;**1**(1):21-50

[93] Rodriguez-Sanjuan JC, Gomez-Ruiz M, Trugeda-Carrera S, Manuel-Palazuelos C, Lopez-Useros A, Gomez-Fleitas M. Laparoscopic and robot-assisted laparoscopic digestive surgery: Present and future directions. World Journal of Gastroenterology. 2016;**22**(6):1975-2004

[94] BenMessaoud C, Kharrazi H, MacDorman KF. Facilitators and barriers to adopting robotic-assisted surgery: Contextualizing the unified theory of acceptance and use of technology. PLoS ONE. 2011;**6**(1):e16395. DOI: 10.1371/journal.pone.0016395

[95] Alemzadeh H, Raman J, Leveson N, Kalbarczyk Z, Iyer RK. Adverse events in robotic surgery: A retrospective study of 14 years of FDA data. PLoS ONE. 2016;**11**(4):e0151470. DOI: 10.1371/journal.pone.0151470

[96] Ielpo B, et al. An organizational model to improve the robotic system among general surgeons. Il Giornale Di Chirurgia. 2014;**35**:52-55

[97] Oviedo RJ, Robertson JC, Alrajhi S. First 101 robotic general surgery cases in a community hospital. JSLS. 2016;**20**(3):e2016.00056. DOI: 10.4293/JSLS.2016.00056

Chapter 4

Bilateral Axillo-Breast Approach Robotic Thyroidectomy: Introduction and Update

Do Hoon Koo, Dong Sik Bae and June Young Choi

Additional information is available at the end of the chapter

http://dx.doi.org/10.5772/intechopen.68951

Abstract

Bilateral axillo-breast approach (BABA) endoscopic thyroidectomy was introduced at Seoul National University Hospital in 2004, and it has been used to treat a variety of benign and malignant thyroid diseases. In 2008, we began using the da Vinci robotic system with BABA endoscopic thyroidectomy and reported our initial experiences in 2009. Since then, the outcomes of many clinical studies have been reported. In this chapter, we will introduce the BABA robotic thyroidectomy (RoT) procedure and review evidence for the safety of performing BABA. First, we will introduce the history of BABA RoT, which is based on an endoscopic BABA method. Second, we will review the BABA RoT equipment, operating room (OR) setup, and the procedures, including surgical indications. Third, technical, oncological, and functional evidence for the safety of performing BABA will be described. Fourth, we will highlight the esthetic superiority of BABA RoT compared with conventional thyroidectomy. Finally, the BABA robotic modified radical neck dissection procedure will be introduced, with mention of our experiences and special concerns. We conclude that BABA RoT is technically, oncologically, and functionally safe. In addition, its esthetic superiority should be emphasized. Further research on the prognosis of patients treated by BABA RoT should follow in the future.

Keywords: bilateral axillo-breast approach, robot, thyroidectomy

1. History and introduction to bilateral axillo-breast approach robotic thyroidectomy (BABA RoT)

1.1. Beyond the endoscopic limits

Thyroid carcinoma is the most common endocrine malignancy. Although the treatment of choice for patients with thyroid carcinoma is conventional open thyroidectomy (OT), it inevitably leaves scarring in the neck because of the anatomical location of the thyroid. Thyroid carcinoma is especially prevalent in young women. The prognosis of thyroid carcinoma is favorable, which increases concerns related to quality of life in terms of postoperative neck scars. To avoid cosmetically unfavorable outcomes, a variety of remote approaches have been used in patients at low risk of recurrence. The two most common techniques are the transaxillary approach (TAA) and bilateral axillo-breast approach (BABA). BABA consists of two axillary incisions 0.8 cm in size and two circumareolar incisions, one left (0.8 cm) and one right (1.2 cm).

BABA endoscopic thyroidectomy is a modification of Axillo Bilateral Breast Approach (ABBA) developed by Shimazu et al. [1]. It was introduced at Seoul National University Hospital (SNUH) in 2004 and has since been used to treat a variety of benign and malignant thyroid diseases. Compared with OT, BABA Endoscopic Thyroidectomy (ET) yields comparable postoperative complication rates and thyroglobulin levels but with excellent cosmetic results [2, 3]. Based on these results, in 2008, we combined our unique BABA thyroidectomy technique with the fundamental advantages of the da Vinci robotic system. These advantages include a good operative view using high-definition three-dimensional imaging, an EndoWrist function that enables a high degree of freedom of motion, a tremor-filtering system, and a short learning curve. This enables precise surgical maneuvers to be performed in difficult and narrow workspaces, even though robotic thyroidectomy (RoT) is associated with some disadvantages including high cost, longer operation time, and lack of tactile sensation. In 2009, we reported our initial experiences with BABA RoT at SNUH, which was the first report of its use [4]. Since then, there have been many clinical studies assessing the surgical outcomes and safety of BABA RoT. In this chapter, we aim to introduce the detailed procedure of BABA RoT and review evidence from published studies regarding the technical, oncological, and functional safety and cosmetic outcomes of BABA RoT.

2. Patient selection: indications and contraindications

The indications for BABA RoT are as follows: (1) a well-differentiated thyroid carcinoma such as papillary or follicular thyroid carcinoma <4 cm in diameter, regardless of preoperative lymph node (LN) involvement, (2) minimal invasion of the anterior thyroid capsule and strap muscle, (3) Graves' disease (recommended for <100 ml in volume), (4) male patients who experienced difficulty with endoscope application, (5) a larger benign thyroid nodule or follicular neoplasm (5–8 cm) not eligible for treatment by a conventional endoscopic approach, and (6) obese patients (body mass index [BMI] >30), who cannot undergo endoscopy [4–10]. Absolute contraindications to RoT include patients with distant metastasis, thyroid malignancies that are likely to recur (e.g., medullary thyroid carcinoma, undifferentiated, or poorly differentiated thyroid carcinoma) and are located posteromedially and thus may be very close to the recurrent laryngeal nerve (RLN) or may have invaded into the tracheal wall and concomitant obvious breast malignancy [11, 12]. Thyroid nodules >8 cm in diameter or those in substernal goiters are also relative contraindications to RoT. BABA thyroid surgery does not involve the breast parenchyma in subcutaneous dissection after circumareolar incision. Consequently, previous breast-conserving surgery due to breast cancer or breast augmentation is not a contraindication to BABA RoT. In addition, Kim et al. recently reported that BABA RoT and lateral LN dissection were performed simultaneously in a thyroid carcinoma patient with preoperative cervical LN metastasis [13]. Therefore, BABA RoT is selectively applicable in patients with suspected lateral LN metastasis.

3. Basic equipment and operating room (OR) setup

3.1. Operating theater

A robotic system requires more space than does either open or endoscopic surgery. Therefore, most hospitals have a dedicated robot operating room. The room is maintained such that surgery can be performed under aseptic conditions.

3.2. da Vinci Si HD surgical system (Intuitive, Sunnyvale, CA, USA)

3.3. Instruments

1. Endoscope: Φ10 mm, 30° endoscope
2. Thyroid pillow (Emtas, Seoul, Korea) (**Figure 1**)

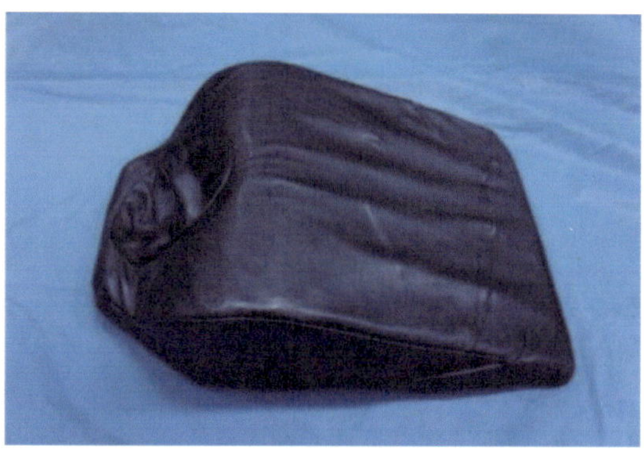

Figure 1. Thyroid pillow.

3. EndoWrist instruments (**Figure 2**)
    - 1 Maryland bipolar forceps, Φ8 mm
    - 1 Prograsp TM forceps, Φ8 mm
    - 1 Cautery hook, Φ8 mm
    - 1 Harmonic®, Φ8 mm

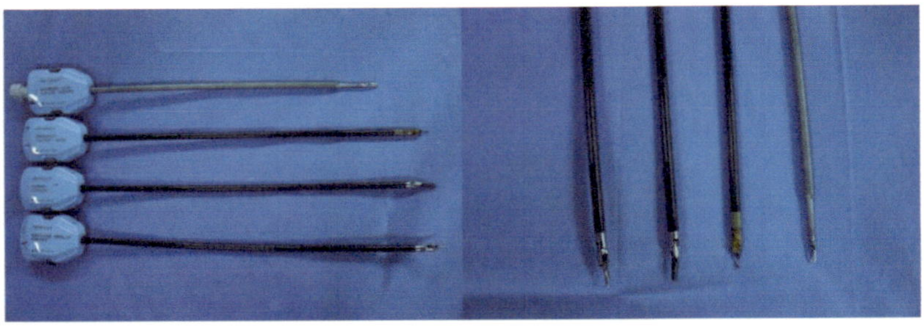

Figure 2. Endowrist instruments.

4. Harmonic® (Ethicon Endo-surgery, Cincinnati, OH, USA)
5. Vascular tunneler (Gore-Tex) (**Figure 3**)

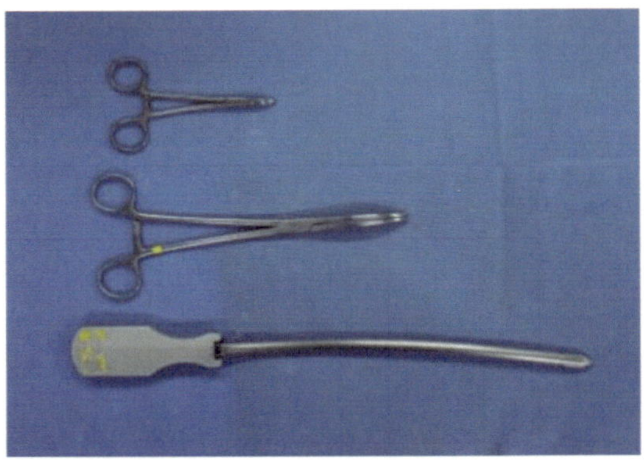

Figure 3. Vascular tunneler.

6. Trocars (**Figure 4**)

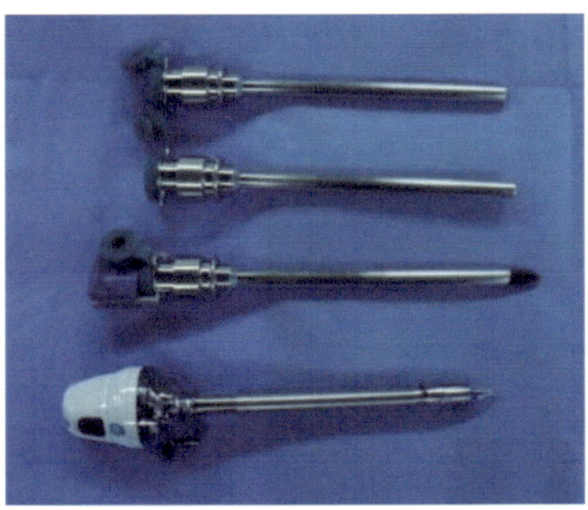

Figure 4. Trocars.

http://dx.doi.org/10.5772/intechopen.68951

7. Endobag, 10 mm (**Figure 5**)

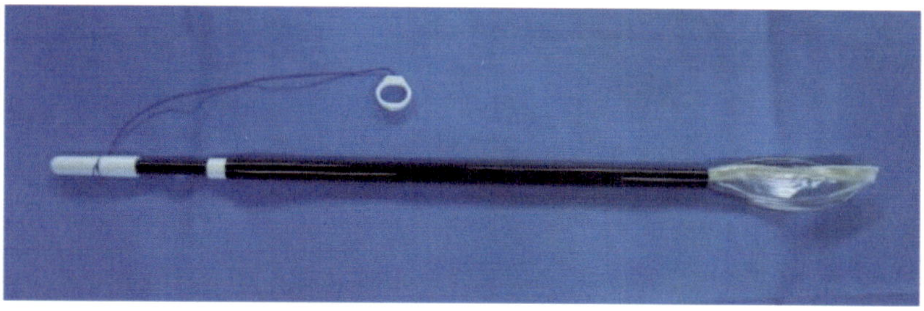

Figure 5. Endobag.

8. Suction-irrigator (**Figure 6**)

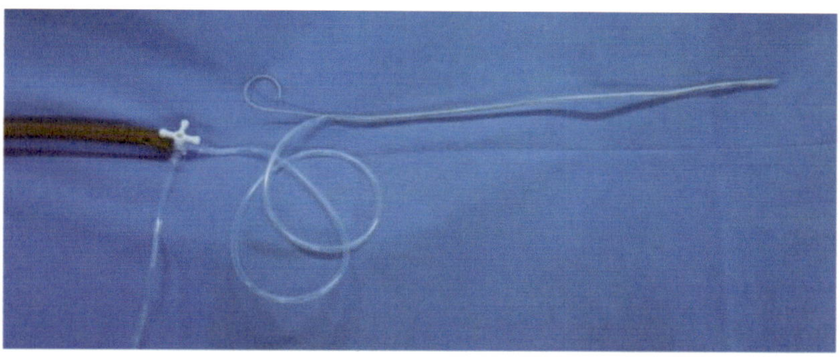

Figure 6. Suction-irrigator.

9. Other instruments (**Figure 7**)

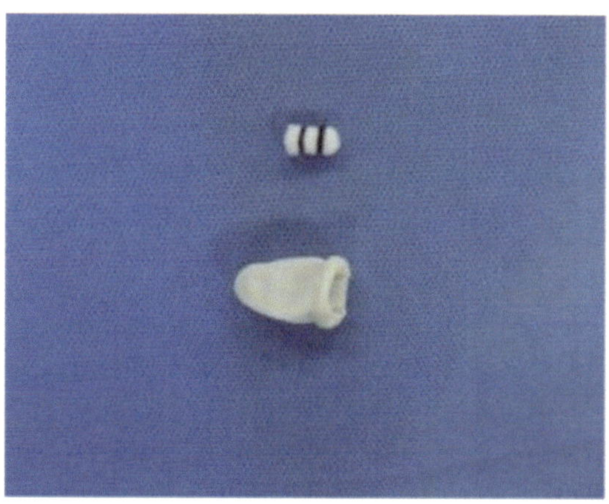

Figure 7. Peanut and thimble.

10. OR Setup (**Figure 8**)

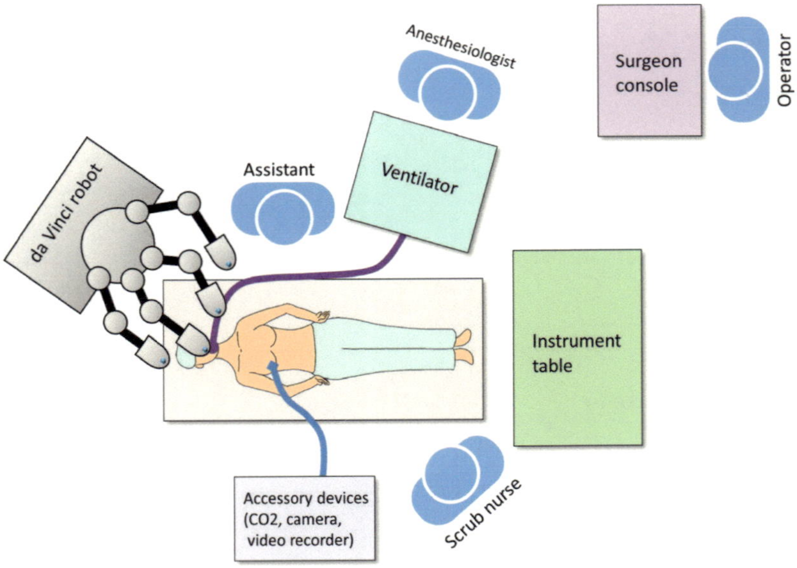

Figure 8. Schematic depiction and the view from above for the operating room setting on robotic thyroidectomy.

4. Procedure and techniques

4.1. Thyroidectomy

4.1.1. Preparation

4.1.1.1. Positioning and draping

Under general anesthesia, the patient is placed in the supine position with a Q-pillow under the shoulder extending the head and neck and the arm resting alongside the body (**Figure 9**).

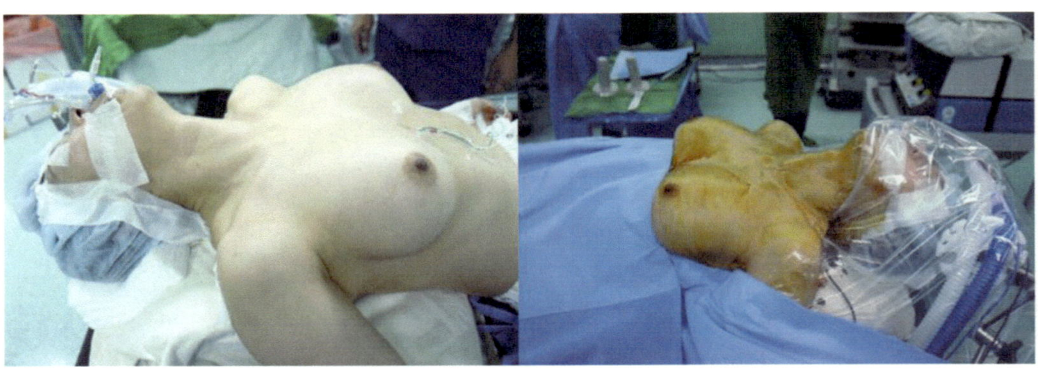

Figure 9. Position and drape.

Care should be taken not to overstretch the patient's neck. Alternatively, Kang et al. suggested a "verticalizing maneuver (VM)" that lifts up the circumareolar sites as high as possible by surrounding the lower part of the lower breast with elastic bands [6]. This method positions the trocar axis more perpendicular, which reduces the blind spot in the lower neck during central compartment node dissection. The surgical field is prepared according to routine surgical maneuvers, and sterile drainage is performed using a universal drape package to expose the anterior neck, bilateral axilla, and lower contour of the breasts. The visual field of the patient's face and endotracheal tube can be maintained by covering the patient's head and face with a transparent plastic sheet (**Table 1**).

1. Preparation
(1) Positioning and draping
(2) Drawing guidelines
(3) Epinephrine-mixed saline injection
2. Flap making
(1) Skin incision and blunt dissection
(2) Port insertion and sharp dissection using an energy device
(3) Robot docking and complete elevation of the flap
3. Thyroidectomy on the lesion side
(1) Midline division
(2) Isthmectomy and/or removal of the pyramidal lobe and midline LN
(3) Lateral and anteromedial dissection of the thyroid gland
(4) Dissection of the thyroid lower pole
(5) Preservation of the recurrent laryngeal nerve and parathyroid glands
(6) Dissection of the thyroid upper pole
4. Specimen removal
5. Central compartment dissection and contralateral thyroidectomy (if indicated)
6. Closure

Table 1. Surgical steps of bilateral axillo-breast approach robotic thyroidectomy.

4.1.1.2. Drawing guidelines

Guidelines are drawn along the following anatomical markings of the chest and neck: thyroid cartilage notch, cricoid cartilage (+), suprasternal notch (U), midline connecting them above, anterior border of the sternocleidomastoid muscle (SCM), superior border of the clavicle and 2 cm below the border, incisions (two circumareolar incisions at the superomedial margins and two axillary incisions using conventional skin wrinkles), and four trajectory lines from each of four skin incision sites to the cricoid cartilage and workspace (**Figure 10**). The dissecting area is bordered by the thyroid cartilage superiorly, 2 cm below the superior border of the clavicle inferiorly and just beyond the medial border of the SCM muscles laterally.

4.1.1.3. Epinephrine-mixed saline injection

Diluted (1:200,000) epinephrine solution is injected into the workspace below the platysma of the neck and subcutaneously into the anterior chest. A 23-G spinal needle is then used to check the intravenous puncture by pulling the syringe back slightly before injecting the solution (**Figure 11a**). At this time, it is possible to inject the solution more securely while avoiding puncturing the blood vessel by bending the needle slightly at an angle. A "pinch and raise"

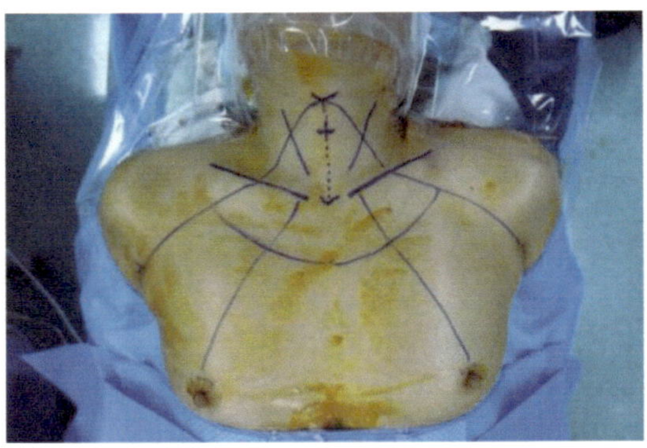

Figure 10. Drawing guideline.

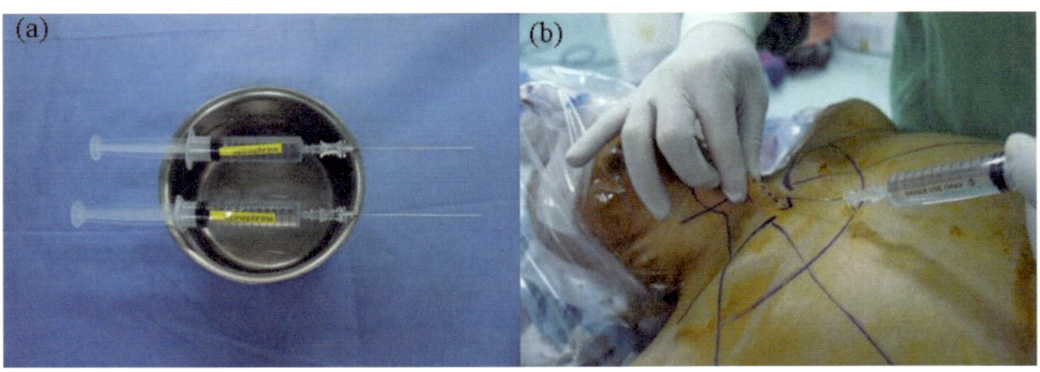

Figure 11. Epinephrine-mixed saline injection (a) 23 G spine needle (b) "pinch and raise" technique.

maneuver of the skin from the neck area facilitates injection of saline into the subplatysmal area (**Figure 11b**). This "hydrodissection" technique is used to create a saline pocket in the subplatysmal layer to reduce bleeding in the flap area and facilitate subsequent dissection. Additionally, Kang et al. previously reported that infiltration of the flap sites with a ropivacaine-saline solution (100 cc normal saline mixed with 3 mg/kg 0.1% ropivacaine) is a safe and effective method for reducing postoperative pain and postoperative analgesic need [14].

4.1.2. Flap making

4.1.2.1. Skin incision and blunt dissection

A circumareolar incision is made along the superomedial margin of each areola (**Figure 12**). First, a 12-mm incision is made on the right side to be used as a camera port, and the subcutaneous tissue is dissected using an electric cauterizer. Next, a straight mosquito hemostat, a long Kelly clamp, and a vascular tunneler are used to generate a subcutaneous narrow tunnel along the trajectory line for trocar insertion. Blunt dissection of the flap formed by hydrodissection begins at zone 2 and extends to zone 1 using a vascular tunneler. At this time, excessive force must not be used when performing blunt dissection near the sternal notch. Next, an 8-mm incision is made on the superomedial margin of the left areola, and blunt dissection of zones 1 and 2 is completed by repeating the same procedure described above.

4.1.2.2. Port insertion and sharp dissection using an energy device

After blunt dissection of the flap from the incision sites to the cricoid cartilage using the tunneler, the ports are inserted through the incision (**Figure 13**). The flap is located higher than the breast parenchyma so that it does not injure the patient's breast. The 12-mm camera port is inserted through the right breast incision, and the 8-mm port is inserted through the left breast incision. At this time, the port insertion sites around the areola are encircled with Duoderm® to cover and protect the areolar after the port is inserted, prevent skin burns on the incision surface, and avoid air leakage. The workspace is maintained at low pressure (5–6 mmHg) by pumping CO_2 gas through the 12-mm camera port [15]. The ultrasonic shear (Harmonic, Ethicon EndoSurgery Inc., Cincinnati, OH, USA) is inserted through the 8-mm port on the left areolar incision to meet the camera through the 12-mm port and to secure the field of view and remove the remaining trabeculae of the subcutaneous tissue. After creating a workspace in the anterior chest (zone 1 and/or 2), avoiding the firm area near the sternal notch, two 8-mm incisions are made, and the trocar is inserted along the axillary trajectory line.

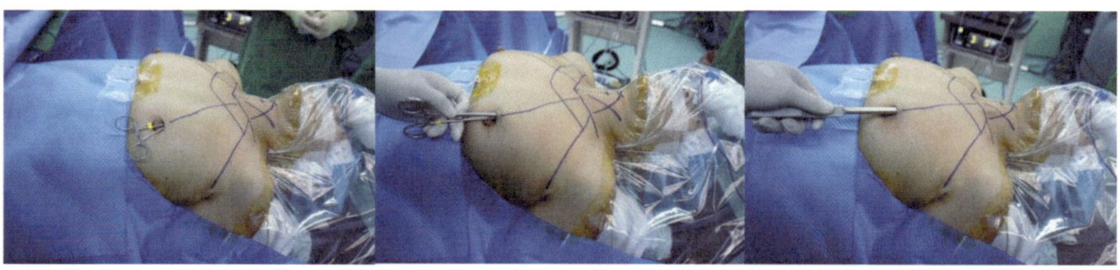

Figure 12. Skin incision and blunt dissection.

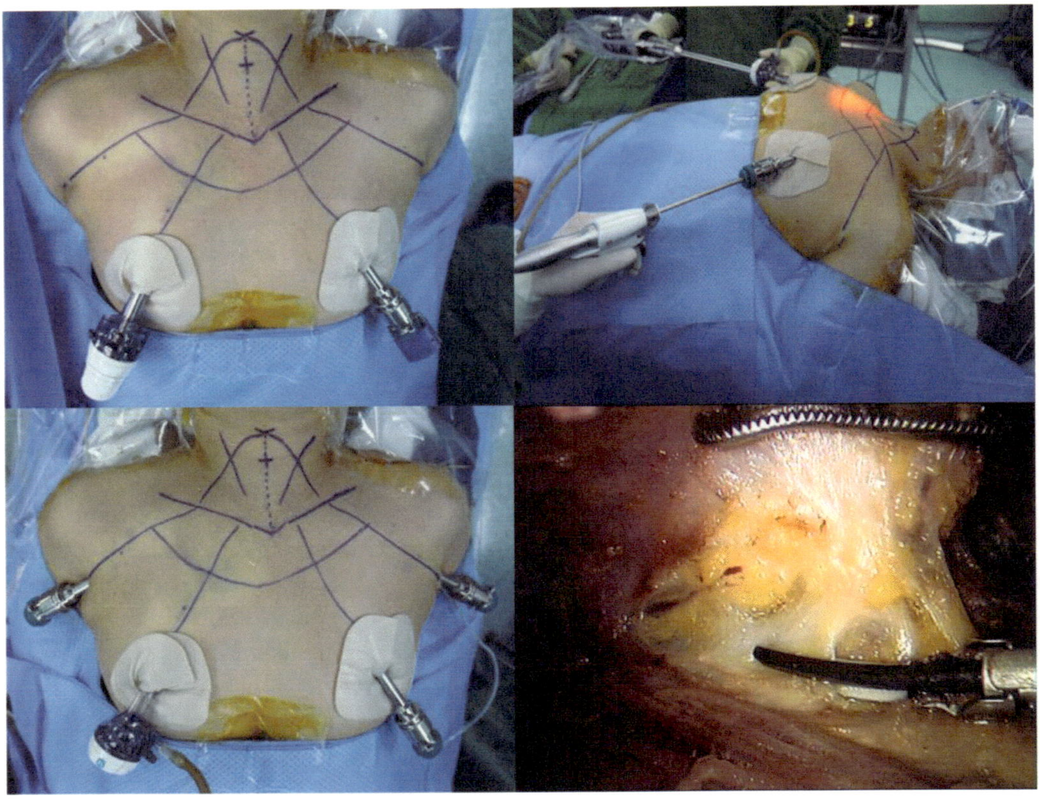

Figure 13. Four ports insertion and sharp dissection with energy device.

4.1.2.3. Robot docking and complete elevation of the flap

After inserting the four ports, the operation bed is switched to a reverse Trendelenburg position of ~20–30°. The central columns of the robot carts and the camera arm are aligned with the camera port in a straight line, and the robot is docked to the port and connected via each of the four robot arms (**Figure 14**). The camera is inserted into the right areolar incision site port, and a monopolar electrocautery or ultrasonic shear is inserted into the left port. Graspers (ProGrasp forceps and Maryland forceps, Intuitive Surgical Inc., Sunnyvale, CA, USA) are inserted through both axillary ports, and further dissection is performed (**Figure 15**). This procedure completes the flap safely and effectively without bleeding. The border of the completed flap extends from the thyroid cartilage superiorly to 2 cm below the clavicle and to the point just beyond the medial margin of the SCM muscle. Recent reports suggested that subfascial layers likely cause less postoperative adhesion than do conventional subplatysmal layers in making flaps [16]. Anterior jugular vein ligation is necessary for dissections performed using the subfascial layer, which can be safely ligated near the sternal notch using an ultrasonic shear or a bipolar coagulator connected to Maryland forceps.

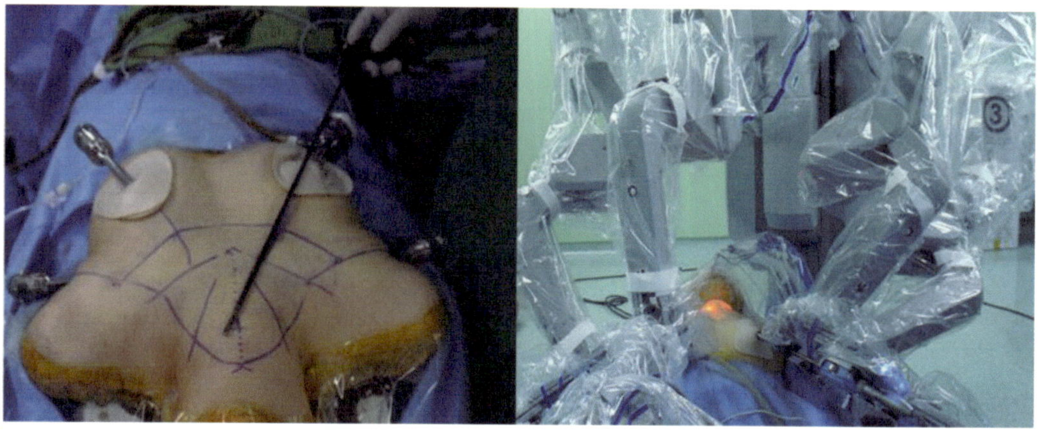

Figure 14. Robot docking.

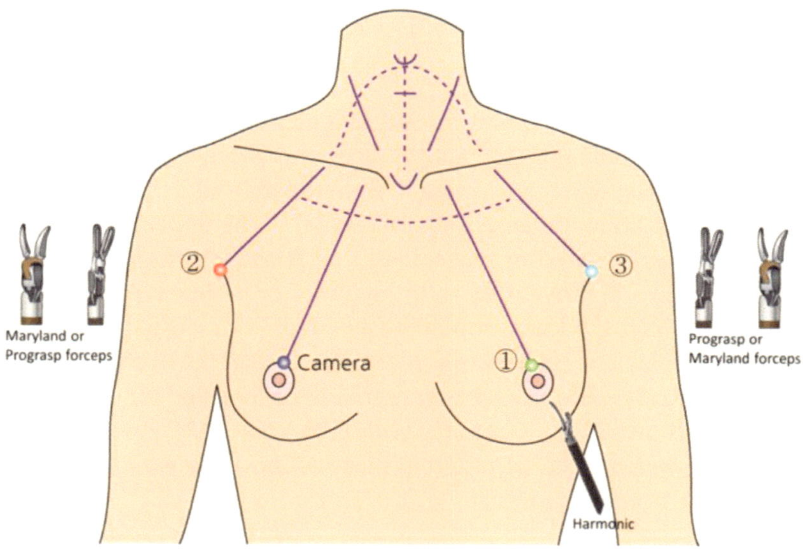

Figure 15. Placement of robotic instruments.

4.1.3. Thyroidectomy on the lesion side

4.1.3.1. Midline division

The first step of BABA RoT is resection, performing a midline division of the strap muscle in a similar fashion to conventional OT (**Figure 16**). The midline between the strap muscles is identified and separated by monopolar electrocautery. At this time, the cervical fascia is opened from the suprasternal notch to the thyroid cartilage to expose the entire length of the strap muscle. For identifying the midline, it is helpful for confirming the boundary that the first assistant palpates the prominence of the thyroid cartilage and the suprasternal notch from the outside.

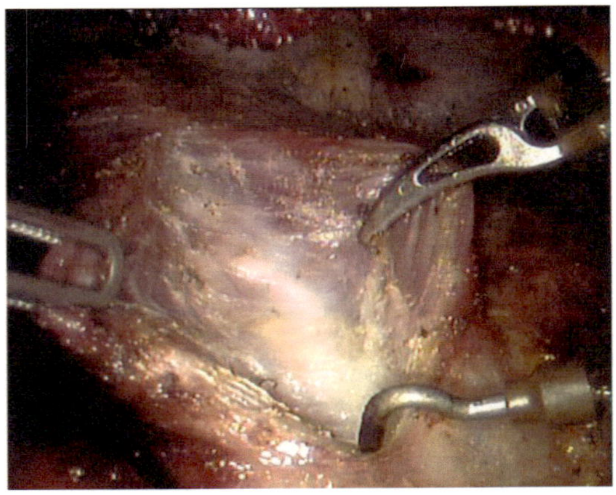

Figure 16. Midline division.

4.1.3.2. Isthmectomy and/or removal of the pyramidal lobe and midline (pretracheal and prelaryngeal) LNs

After verifying the trachea, isthmus, and cricothyroid membranes in the visual field, the isthmus is separated by ultrasonic shear or hook electrocautery (**Figure 17**). The trachea is easily identified by dissecting the soft tissue caudally from the thyroid isthmus, taking care not to injure the trachea. In addition, because there is a vessel in the upper border of the isthmus, care should be taken to avoid bleeding when dissecting. It is important to confirm the presence of isthmus lesions on preoperative images. If the tumor or nodule is located in the isthmus on the preoperative image, the lesion should be avoided, i.e., by using the paraisthmic line. Sometimes, the pyramidal lobes extend cranially to the level of the hyoid bone, and a thyroid duct cyst is detected incidentally. This structure should be removed for complete resection of the thyroid tissue; this procedure is possible with BABA RoT [17]. Furthermore, a delphian or prelaryngeal node between the cricothyroid muscles above the isthmus and a pretracheal

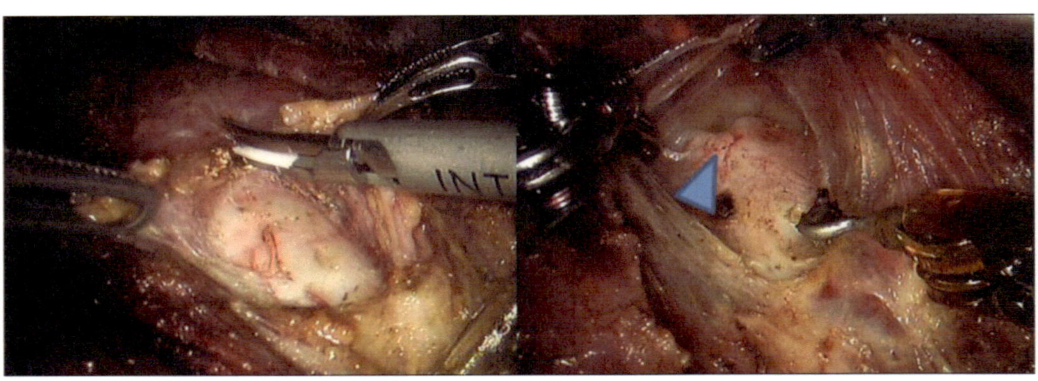

Figure 17. Isthmectomy, arrow; thyroid notch.

node below the isthmus shoulder may be found during soft tissue dissection. If LN metastasis is suspected, it is possible to excise the LN and confirm metastasis intracorporeally using frozen biopsy [18]. This area always contains small blood vessels, but monopolar electrocautery allows hemostasis. Extra attention is needed to avoid injuring the cricothyroid muscles during dissection.

4.1.3.3. Lateral and anteromedial dissection of the thyroid gland

After isthmectomy and/or midline LN resection, the thyroid gland on the lesion side is retracted medially using ProGrasp forceps, and the strap muscle is retracted laterally using Maryland forceps to separate the strap muscle from the capsule of the thyroid gland (**Figure 18**). This dissection extends to the deep aspect of the gland to expose the lateral side of the thyroid gland. Upon lateral dissection, the middle thyroid vein is visible and is ligated using ultrasonic shears or Maryland forceps. Ultrasonic shears are useful to reduce unnecessary bleeding from the muscles and thyroid capsule during this process. The so-called "switching action," which moves the thyroid gland in the medial direction in phase with two robotic arms, facilitates medial retraction of the thyroid gland. In addition, the thyroidectomy procedure may be facilitated by dissection of the medial side (peritracheal and cricoid cartilage) as well as the lateral side. Further dissection is then performed from the lower pole to the medial side of the trachea in accordance with the principle of capsular dissection.

4.1.3.4. Dissection of the thyroid lower pole

After completing the lateral and medial dissections of the thyroid gland, the next step is dissection of the inferior portion of the thyroid gland (**Figure 19**). The lower pole of the thyroid gland is dissected bluntly using ultrasound scissors or Maryland forceps, because the inferior thyroid artery passes directly below or crosses over the recurrent laryngeal nerve before entering the thyroid gland. Therefore, the inferior thyroid artery can be used as an anatomical guide for exposing the recurrent laryngeal nerve.

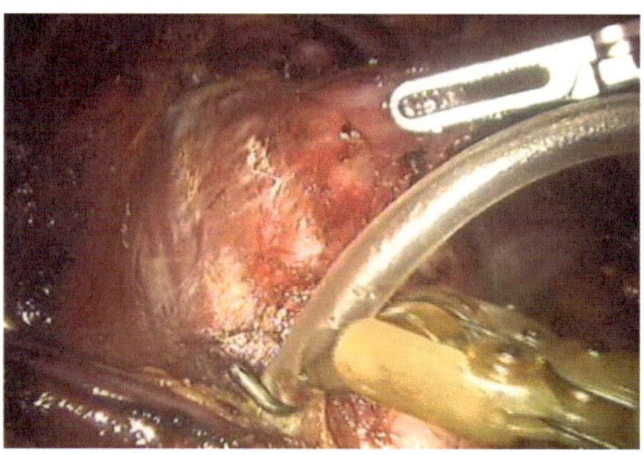

Figure 18. Lateral dissection of the thyroid gland.

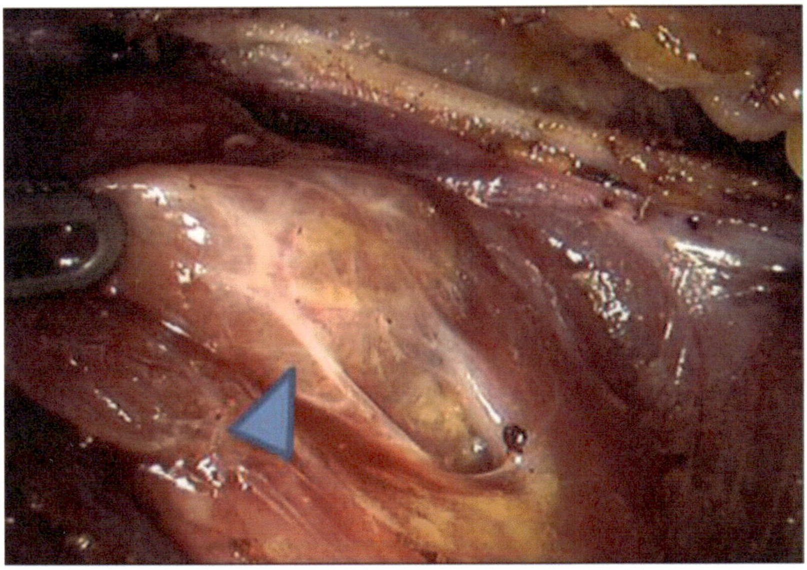

Figure 19. Dissection of the thyroid lower pole, arrow; inferior thyroid vein.

4.1.3.5. Preservation of the RLN and parathyroid gland (PTG)

During dissection of the thyroid gland from the perithyroidal tissue, it is important to preserve the RLN and PTG (**Figure 20**). The RLN and PTG should be identified while carefully dissecting the inferolateral side of the thyroid gland. Once the RLN is found, a plane delineated just superficial to the nerve and the ligament of Berry is separated using ultrasonic shears. Dissection progresses in the cephalad direction to the point where the nerve enters the larynx. Near the ligament of Berry, careful dissection is needed to avoid traction or thermal injury to the RLN. It was reported that intraoperative neuromonitoring can help identify and preserve the RLN [19, 20]. In addition, Yu et al. introduced near-infrared light-induced indocyanine green fluorescence to identify the PTG during BABA RoT and reduce the risk of

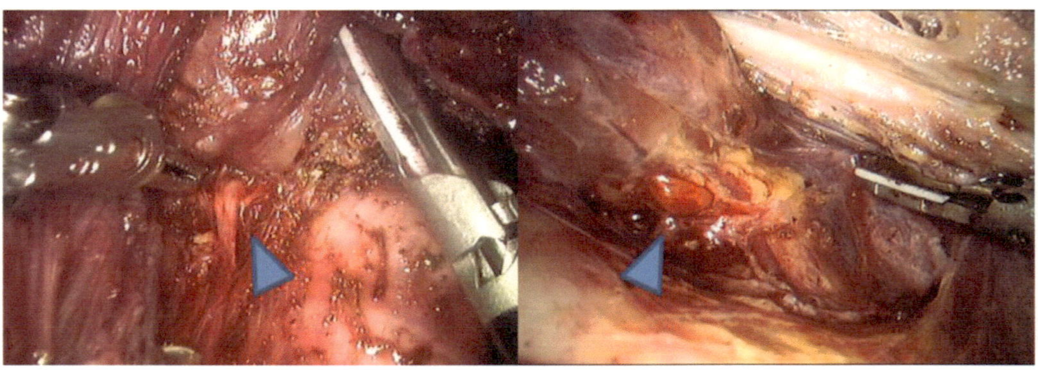

Figure 20. Preservation of the recurrent laryngeal nerve and parathyroid gland.

http://dx.doi.org/10.5772/intechopen.68951

incidental parathyroidectomy [21]. If the nerve is not immediately exposed, the loose fibrous tissue needs to be further dissected from the inferior point of the artery near the tracheo-esophageal groove. At this time, the inferior PTG, which can be used as a guide to the RLN, can be detected. The Zuckerkandl tubercle can also be used as a guide to the RLN. Therefore, the area under the Zuckerkandl tubercle requires caution when dissecting using Maryland forceps. Because the inferior thyroid vessels supply blood to the inferior PTGs, the inferior vessels should be ligated close to the thyroid to preserve blood flow. If preservation of the PTGs is not possible, reimplantation should be considered. The pectoralis major muscle is preferred for autotransplantation of the PTG.

4.1.3.6. Dissection of the thyroid upper pole

With the retractor pulling the upper portion of the strap muscles in a cephalad direction and the trachea in a medial direction, ultrasonic shears are used to dissect the upper pole of the thyroid gland (**Figure 21**). The medial and lateral sides are dissected alternately to separate the upper pole of the thyroid gland. It is important to preserve the fascia of the cricothyroid muscle, because the external branch of the superior laryngeal nerve is closely related to the cricothyroid muscles [22]. Therefore, it is helpful to maintain the fascia using medial traction of the trachea during this procedure. In most cases, the posterior branch of the upper thyroid vessel, which supplies blood to the superior PTG, can be preserved by careful capsular dissection. There may be one or two small veins entering the posterior portion of the upper pole; these vessels should be identified and ligated carefully. Then, the terminal branches of the superior thyroid artery and vein should be identified and ligated carefully using ultrasonic shears. The three approaches to dissecting the upper thyroid gland are the (1) lateral, (2) anteromedial, and (3) posterior medial approaches. The lateral approach refers to gradual and careful dissection of the strap muscles attached to the thyroid gland. The anteromedial approach to the thyroid upper pole corresponds to extending the space between the thyroid gland and the anterior portion of the cricothyroid muscle. The posteromedial approach involves coming in close contact with the superior thyroid vessels along the ligament of Berry and cricothyroid fascia.

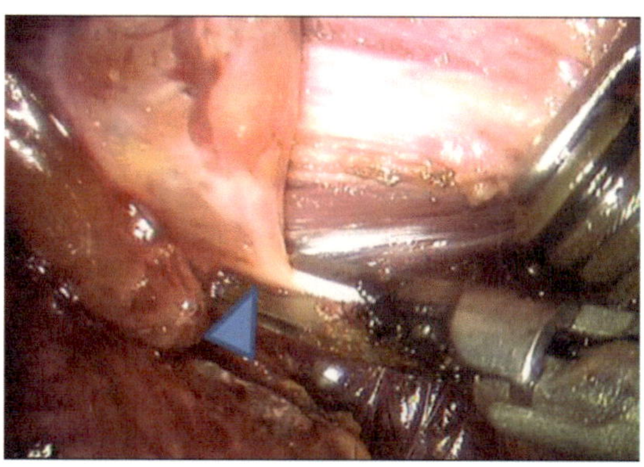

Figure 21. Dissection of thyroid upper pole, arrow, superior thyroid artery.

4.1.4. Specimen removal

After complete dissection of the thyroid gland from the trachea, the specimen is wrapped in an endoplastic bag (LapBag; Sejong Medical, Seoul, Korea) and removed through the left axillary port (**Figure 22**). If the incision of the left axilla is insufficient to extract the specimen, the incision can be widened using a knife. Once the specimen is extracted, it is diagnosed by analyzing intraoperative frozen sections and used to determine the extent of the operation required.

4.1.5. Central compartment dissection and contralateral thyroidectomy (if indicated)

If the frozen section is confirmed as malignant, central LN dissection (therapeutic or prophylactic) should be performed (**Figure 23**). Care should be taken to avoid injury to the recurrent laryngeal nerve by central compartment dissection. The contralateral lobe is handled in the same way. As shown in the figure, the operator has a comfortable and symmetrical view of the surgical field using BABA.

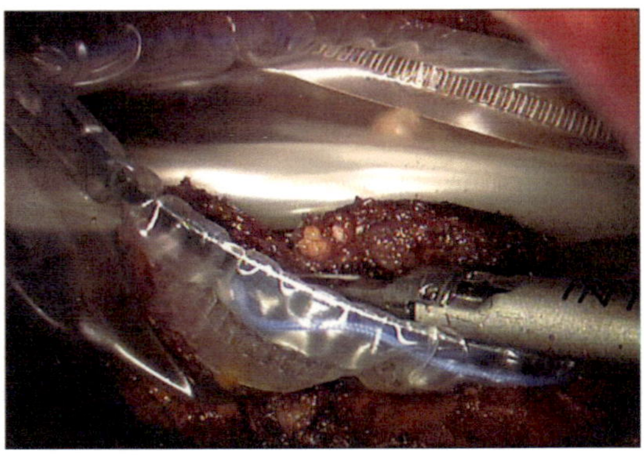

Figure 22. Specimen removal using endobag.

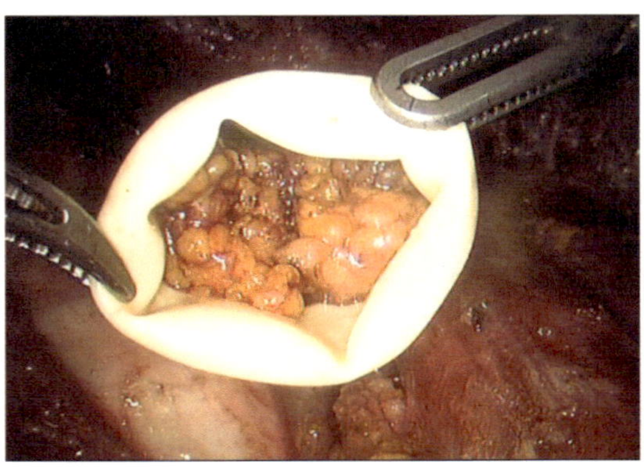

Figure 23. Central lymph node removal using thimble.

4.1.6. *Closure*

After the thyroidectomy is completed, the operative field is irrigated with warm saline. Hemostasis is performed carefully, and fibrin sealant (Tisseel®; Baxter Healthcare Corporation, Westlake Village, CA, USA) is then applied if necessary. The antiadhesive material is placed between the trachea and strap muscle and then between the skin and fascia. The midline between the two strap muscles is closed by a continuous running suture (**Figure 24**). Then, one or two Jackson-Pratt drains are inserted into the thyroid pockets through the opposite or bilateral axillary incisions; however, drainless BABA thyroidectomy was reported to be feasible [23]. It was also reported that a ropivacaine solution can be instilled into the skin flap before skin closure to reduce postoperative pain and the requirement for analgesia [24]. Finally, the skin of both breasts and the axilla are sutured by the knot-burying technique using an absorbable ligature.

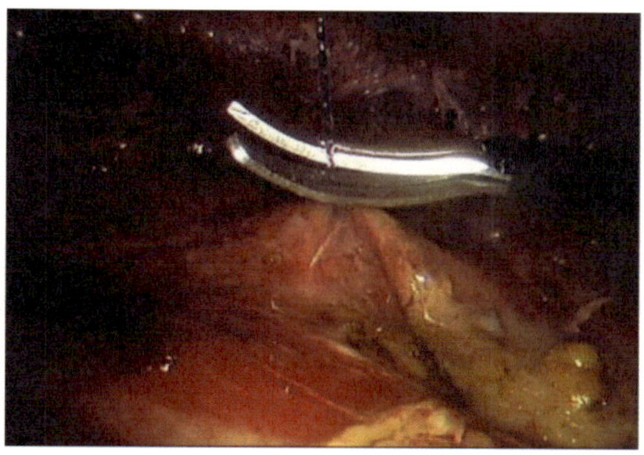

Figure 24. Midline closure.

4.2. LN dissection

4.2.1. *Central compartment dissection* (**Figure 25**)

After completion of thyroidectomy on the lesion side, ipsilateral neck LN dissection is performed. For therapeutic central LN dissection, it is particularly important to avoid RLN injury, preserve the PTG, and achieve complete resection of the suspected LN. In advance, it is useful to have a spacious field of vision to expose the central LNs and major structures. Kang et al. reported that blind spots are reduced using a deep-seated LN approach around the central compartment below the sternal notch via a VM that repositions the pivot point of the robot arm as high as possible [6]. In addition, Kim et al. reported that the addition of a snake retractor to the axillary trocar site enhances the central view and increases the number of resected LNs [23].

For complete and safe central LN dissection, an understanding of the anatomical relationship among the thymus, lower PTG, and soft tissues containing the LNs is needed. The vertical inferior thyroid veins running along the thymus help to indicate the dissection plane. The central compartment LN is located deeper vertically than the plane of these veins and the thymus.

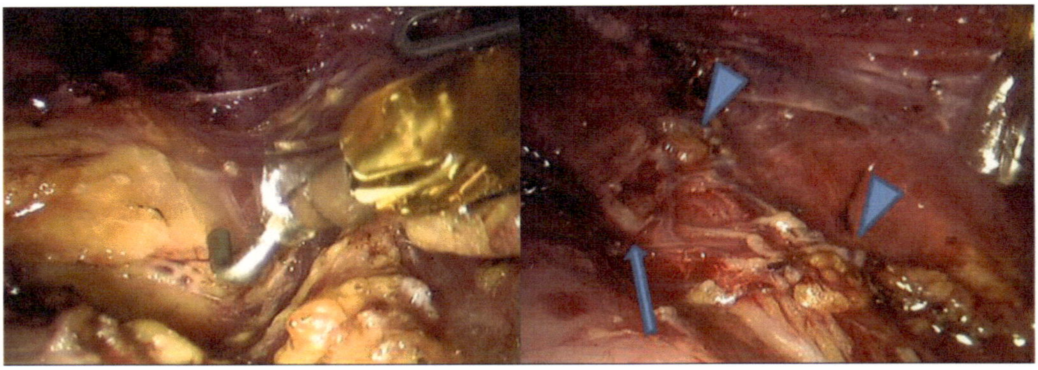

Figure 25. Central compartment dissection, arrow long; left recurrent laryngeal nerve, arrow short; left superior and inferior parathyroid glands.

The inferior PTG is located on the superficial plane, usually within or near the thymus. Thus, preserving the thymus helps reduce the risk of hypoparathyroidism (hypoPTH). Usually, the central LN is separated, with preservation of the thymus and the inferior PTG, and removed from the carotid artery in the medial direction. If a PTG is accidentally removed along with the resected tissue containing the central LN, autotransplantation into the pectoralis major is recommended. The RLN should be carefully monitored and preserved at this stage. Therefore, a nerve-monitoring device connected to a monopolar electrocautery is helpful for identifying the RLN.

4.2.2. Lateral compartment dissection

The procedures related to BABA robotic lateral neck dissection are essentially similar to those of the open method and have been reported previously [13]. First, this procedure requires a larger skin flap than that required for conventional thyroidectomy, with the boundaries being the inferior border of the submandibular glands cranially, the mandible angle superiorly, and the anterior edge of the trapezius muscle posteriorly. The fascia between the sternothyroid muscles and the SCM muscles is incised. After the medial and lateral borders of the SCM muscle are fully exposed, the SCM muscle is pulled upward using a #0 polydioxanone suture (Ethicon, San Angelo, TX, USA) and fixed. In the level IV dissection, the transverse cervical artery and phrenic nerve are identified, and the level II dissection is extended until the posterior belly of the digastric muscle preserves the spinal accessory nerve. The direction of the camera port can be changed such that the dissecting field of view is secured and pulled further cranially when necessary. It can also be helpful to rotate the camera port slightly clockwise or counterclockwise.

5. Review of the evidence: the safety of performing BABA RoT

5.1. Technical

Table 2 shows the technical safety parameters for BABA RoT. Below, we describe various surgical complications, including RLN paralysis and hypoPTH, the most important factors for thyroidectomy.

First author, year	No. of samples (total cases)	VC palsy*	VC palsy¶	HypoPTH*,‡	HypoPTH¶,‡	Bleeding	Chyle leak
Kim, 2011 [25]‡	69 (69)	1.4%	0%	33.3%	1.4%	0%	1.4%
Lee, 2013 [8]	1026 (872)	14.2%‡	0.2%‡	39.1%	1.5%	0.4%	NA
Kim, 2014 [6]	123 (100)	4.9%	0%	29%	0%	0%	NA
Lee, 2015 [26]	100 (88)	3.0%	0%	21.6%	0%	0%	0%
Kim, 2015 [23]	300 (143)	2.6%	0%	23.1%	1.4%	0.3%	0.6%
Cho, 2016 [27]	109 (99)	6.4%	0.9%	33.0%	1.8%	0.9%	0%
Bae, 2016 [28]	118 (91)	3.3%‡	0%‡	35.2%	2.2%	0%	NA

*Transient.

¶Permanent.

‡For total thyroidectomy cases.

Note: No.: number; VC: vocal cord; hypoPTH: hypoparathyroidism; NA: not available.

Table 2. Technical safety of performing bilateral axillo-breast approach robotic thyroidectomy.

5.1.1. Recurrent laryngeal nerve (RLN)

Table 2 shows the incidences reported to date of transient and permanent RLN injury during BABA RoT [6, 8, 23, 25–28]. These studies were published in Korea. In most studies, transient RLN damage was defined as hoarseness or vocal fold paralysis of <6 months. The reported incidence of transient RLN injury in patients undergoing BABA RoT ranges from 1.4 to 14.2%, and most studies have reported an incidence of <7%. Particularly, permanent RLN injuries were observed in <1% of patients, which is an excellent result, comparable to that of conventional OT. **Table 3** shows the results of five studies that compared RoT with OT or ET;

First author, year	No. of pts. (RoT vs. OT)	No. of TT (RoT vs. OT)	Evaluation	Transient palsy (%) (RoT vs. OT)	Permanent palsy (%) (RoT vs. OT)
Kim, 2011 [25]‡	69 vs. 138	69 vs. 138	Laryngoscopy	1.4 vs. 0.7 (0.615)	0 vs. 0 (1.000)
Kim, 2014 [6]	123 vs. 392	100 vs. 364	Laryngoscopy	4.9 vs. 6.1(0.607)	0 vs. 0.3 (1.000)
Kwak, 2015 [29]	206 vs. 634	157 vs. 544	Stroboscopy	0.5 vs. 0.9 (0.363)	NA
Cho, 2016 [27]	109 vs. 109*	–	Laryngoscopy	6.4 vs. 5.5 (0.775)	0.9 vs. 0.9 (1.000)
First author, year	**No. of pts. (RoT vs. ET)**	**No. of TT (RoT vs. ET)**	**Evaluation**	**Transient palsy (%) (RoT vs. ET)**	**Permanent palsy (%) (RoT vs. ET)**
Kim, 2011 [25]‡	69 vs. 95	69 vs. 95	Laryngoscopy	1.4 vs. 2.1 (0.757)	0 vs. 2.1 (0.623)
Kim, 2016 [30]	289 vs. 289*	114 vs. 114	Medical record ± laryngoscopy	4.5 vs. 3.8 (0.677)	0.7 vs. 0.3 (1.000)

*After propensity score matching.

‡For total thyroidectomy cases.

Notes: No.: number; pts: patients; NA: not available; TT: total thyroidectomy.

Table 3. Comparison of recurrent laryngeal nerve palsy between bilateral axillo-breast approach robotic thyroidectomy (RoT) and open thyroidectomy (OT) or endoscopic thyroidectomy (ET).

these studies reported no difference in incidence between transient and permanent RLN injuries [6, 25, 27, 29, 30]. Therefore, the technical safety of BABA RoT for RLN preservation has been demonstrated sufficiently. These results were also validated in several meta-analyses of studies that included BABA and TAA methods, with the exception of one study [31–35].

5.1.2. Hypoparathyroidism (hypoPTH)

Table 2 shows the incidence of transient and permanent hypoPTH after BABA RoT [6, 8, 23, 25–28]. The definition of hypoPTH varies but is generally defined according to parathyroid hormone and calcium levels and hypocalcemic symptoms. In most studies, permanent hypoPTH was defined as the need for medication for at least 6 months. The incidence of transient hypoPTH in patients undergoing BABA RoT was 22–39%, and the incidence of permanent hypoPTH in patients undergoing the total thyroidectomy was <3%. This is an important indicator of the technical safety of BABA RoT, which is comparable to traditional OT. Furthermore, in five studies that compared RoT and OT (**Table 4**), the incidence of transient or permanent hypoPTH was similar between RoT and OT, suggesting that BABA RoT is a more appropriate method for total thyroidectomy [6, 25, 27, 29, 30].

5.1.3. Other complications

Among the other complications, bleeding and chyle leak are described in **Table 2**. Bleeding was reported in four out of seven studies with no cases and in the remaining three studies

First author, year	No. of pts. (RoT vs. OT)	No. of TT (RoT vs. OT)	Definition of transient hypoPTH	Transient (%) (RoT vs. OT)	Permanent (%) (RoT vs. OT)
Kim, 2011 [25]	69 vs. 138	69 vs. 138	PTH normalized within 6 months	33.3 vs. 27.5 (0.484)	1.4 vs. 2.9 (0.873)
Kim, 2014 [6]	123 vs. 392	100 vs. 364	Serum calcium <4.0 mEq/L	29.0 vs. 22.0 (0.161)	0 vs. 0 (0.000)
Kwak, 2015† [29]	206 vs. 634	157 vs. 544	iCa <4.4 mg/dL or PTH < 8 pg/mL	14.6 vs. 15.0 (0.296)	NA
Cho, 2016† [27]	109 vs. 109*	–	PTH <13 pg/mL	33.0 vs. 26.6 (0.374)	1.8 vs. 1.8 (1.000)
First author, year	**No. of pts. (RoT vs. ET)**	**No. of TT (RoT vs. ET)**	**Definition of transient hypoPTH**	**Transient (%) (RoT vs. ET)**	**Permanent (%) (RoT vs. ET)**
Kim, 2011 [25]	69 vs. 95	69 vs. 95	PTH normalized within 6 mo	33.3 vs. 25.3 (0.340)	1.4 vs. 3.2 (0.851)
Kim, 2016† [30]	289 vs. 289*	114 vs. 114	PTH <5 pg/mL	38.6 vs. 33.3 (0.408)	0.9 vs. 1.8 (1.000)

*After propensity score matching.

†Including lobectomy cases.

Notes: No.: number; pts: patients; TT: total thyroidectomy; NA: not available; iCa: ionized calcium.

Table 4. Comparison of hypoparathyroidism between bilateral axillo-breast approach robotic thyroidectomy (BABA RoT) and open thyroidectomy (OT) or endoscopic thyroidectomy (ET).

<1% [6, 8, 23, 25–28]. The incidence of chyle leak was low in the two studies that reported this complication (1.4 and 0.6%, respectively) [23, 25]. Postoperative bleeding and hematoma are potentially fatal complications of thyroidectomy, because reoperation may be necessary to resolve the airway compression caused by hematoma. Otherwise, unlike TAA, brachial plexus and tracheal injury have not been reported in BABA RoT [36].

5.2. Oncological safety

The clinical parameters used to assess oncological safety after thyroidectomy include the number of retrieved LNs in the neck, stimulated thyroglobulin (sTg) level, and radioactive iodine (RAI) uptake on whole-body scan (WBS). Both the sTg level and RAI uptake reflect the surgical completeness of thyroidectomy.

5.2.1. LN retrieval

As the main indication of BABA RoT, papillary thyroid carcinoma frequently exhibits locoregional metastasis into the surrounding cervical LNs. Therefore, LN dissection is performed for therapeutic or prophylactic purposes in most institutions, and the number of resected LNs is an indicator of the oncological safety associated with RoT [37, 38]. In all previous studies except for Kim et al. [25], the number of central neck LNs retrieved by RoT was statistically lower than that by OT [6, 23, 27, 29]. Nevertheless, the total LN count was five to nine, which is considered to exceed the minimum level of adequacy for LN dissection in the central compartment (**Table 5**). Only one study has compared RoT with OT in terms of the number of LNs excised during BABA robotic lateral neck dissection for locally advanced cancer, but no significant difference was observed (RoT vs. OT; 12.8 vs. 12.7 LNs) [13]. However, the currently available data indicate that BABA RoT is not superior to OT in terms of the number of central LNs retrieved.

5.2.2. Surgical completeness: sTg level and WBS

The surgical completeness of resection in thyroid carcinoma is generally assessed by measurements of serum thyroglobulin levels after RAI ablation and RAI uptake on posttherapeutic WBS [46, 47]. sTg levels are measured prior to RAI ablation combined with elevated thyroid stimulating hormone (TSH) treatment, via either thyroid hormone withdrawal or recombinant human TSH injection. Increased sTg levels after total thyroidectomy suggest the presence of remnant thyroid tissue. Therefore, a low sTg level is a reliable surrogate marker for the amount of remnant thyroid tissue after total thyroidectomy. **Table 5** shows the results of studies that measured sTg levels after the first RAI ablation following RoT or OT. Five studies reported no statistically significant difference in sTg levels between RoT and OT [6, 23, 25, 27, 39] Compared with the sTg levels (mean, 4.9–10.2; median, 3.8) following OT, endoscopic surgery, or TAA [41, 42, 44, 45], the mean (0.8–1.4) and median sTg levels (0.2–0.6) following BABA RoT were remarkably lower [6, 23, 25, 27, 28, 39]. In addition, the proportion of patients with a sTg level <1.0 ng/mL was much higher: 65–87% after BABA RoT [6, 23, 39] compared with 21–48% after other approaches [40–43] (reported in previous studies). In two meta-analyses performed by Wang et al. and Son et al., there was no statistically significant difference between RoT and OT in terms of sTg levels [33, 35]. However, in another meta-analysis performed by Lang et al., sTg

First author, year	No. cases. (RoT vs. OT)	LN number (RoT vs. OT)	No. RAI cases (RoT vs. OT)	sTg after 1st RAI ablation (RoT vs. OT)	Proportion of cases with sTg <1.0 ng/mL (RoT vs. OT)
Kim, 2011 [25]	69 vs. 138	4.7 vs. 4.8 (0.802)	–	0.8 vs. 0.8 (0.978)	NA
Lee, 2011 [39]	174 vs. 237	NA	174 vs. 237	1.4 vs. 1.2 (0.998)	69.1% vs. 68.6% (0.924)
Kim, 2014 [6]	123 vs. 392	8.7 vs. 10.4 (0.006)	37 vs. 148	1.4 vs. 1.2 (0.652)	75.7% vs. 76.4% (0.931)
Kim, 2015 [23]	300 vs. 300	6.7 vs. 8.9 (<0.001)	68 vs. 130	0.8 vs. 1.8 (0.001)	86.6% vs. 67.6% (0.004)
Kwak,2015 [29]	206 vs. 634	5.9 vs. 8.4 (0.001)	–	NA	NA
Cho, 2016 [27]	126 vs. 689	3.6 vs. 5.1 (<0.001)	67 vs. 52	0.25 vs. 0.2* (0.954)	NA
Bae, 2016 [28]	118 (RoT)		67	0.6*	
References about RAI ablation					
First author, Year	**No. cases (RoT/vs. OT)**	**Approach**		**sTg after 1st RAI ablation (RoT vs. OT)**	**Proportion of cases with sTg < 1.0 ng/mL**
Schlumberger, 2012 [40]	652	Conventional open		–	48.3%
Mallick, 2012 [41]	110	Conventional open		3.8*	21%†
Lombardi, 2007 [42]	152	Minimally invasive video-assisted		5.5	21%
Choi, 2012 [43]	99	Endoscopic BABA		NA	40.3%
Tae, 2014 [44]	62 vs. 183	Gasless unilateral axillo-breast		10.2 vs. 3.9 (<0.001)	NA
Lee, 2014 [45]	43 vs. 51	Transaxillary		4.9 vs. 4.2 (0.674)	NA

*Median.

†sTg < 2.0 ng/mL.

Notes: NA: not available; RAI: radioactive iodine; No.: number; LN: lymph node; sTg: stimulated thyroglobulin.

Table 5. Oncological safety: comparison of surgical completeness between bilateral axillo-breast approach robotic thyroidectomy (BABA RoT) and open thyroidectomy (OT).

levels were significantly higher after robotic compared with open surgery, which was more pronounced after TAA compared with BABA RoT [48].

Remnant thyroid tissue can also be measured by RAI thyroid uptake on WBS. Lee et al. reported that RAI uptake on the initial WBS was similar in the BABA RoT and OT groups after propensity score matching (the two groups were matched using a total of eight factors, including three demographic and five pathological characteristics) to minimize selective bias [39]. This study is the first report to systematically analyze the surgical completeness of BABA RoT and OT. Statistical techniques were applied to improve comparison of the two groups; therefore, this was a meaningful attempt to overcome the limits of a retrospective study design.

5.3. Functional

5.3.1. Pain

Since RoT requires formation of a larger skin flap than that does OT, there is concern that the postoperative neck and chest pain will be greater after RoT. In a prospective study, Chai et al. reported no significant difference in the postoperative pain score for the throat, anterior neck, posterior neck, or back at 1, 2, 3, and 14 days postoperatively between the BABA RoT ($n = 27$) and OT ($n = 27$) groups [49]. They also reported that the postoperative analgesic requirements were similar between the two groups using applications on mobile devices such as the iPAD to facilitate the assessment and management of pain in postoperative patients. In addition, Cho et al. reported similar pain scores between the RoT and OT groups ($P = 0.669$ after surgery, $P = 0.952$ on postoperative day 1) [27]. Koo et al. reported no significant difference in chronic pain levels ($P = 0.321$) between the BABA RoT and OT groups after correcting for age and the postoperative follow-up period [50]. Because previous studies have used different assessment scales, it is not possible to provide a standardized comparison of postoperative pain by meta-analyses; nevertheless, the overall evidence suggests that RoT and OT achieve similar results in terms of postoperative pain. **Table 6** shows the results of three prospective randomized controlled trials (including 108, 55, and 34 subjects, respectively) that attempted to reduce postoperative pain after BABA RoT [14, 24, 51]. In all three studies, the preoperative and postoperative instillation of analgesics (ropivacaine or levobupivacaine spray) to the flap site during BABA RoT reduced postoperative pain and the need for analgesics compared with the OT group.

5.3.2. Voice quality and swallowing function

Postoperative voice quality after BABA RoT, independent of RLN injury, has been assessed in two studies. In 2015, Bae et al. assessed the VHI-10 score before surgery and 2 weeks, 3 months, and 6 months postoperatively [28]. After adjusting for the effect of time, they concluded that the mean Korean VHI-10 score during the postoperative 6 months increased initially but tended to

Author, year	**Study design**	**Number subjects (patients vs. controls)**	**Analgesic**	**Parameters**
Bae et al. 2015 [24]	PRCT	108 (54 vs. 54)	Ropivacaine, postoperative	VAS score, analgesic requirements, and adverse events
Ryu et al. 2015 [51]	PRCT	55 (28 vs. 27)	Levobupivacaine spray, postoperative	Pain score, need for PCA, other adverse effects
Kang et al. 2015 [14]	PRCT, double-blind	34 (17 vs. 17)	Ropivacaine, preincision	VAS score, bottom hit counts from PCA, need for fentanyl, CRP levels, BP, and HR

Notes: PRCT, prospective randomized controlled trial; VAS, visual analog scale; PCA, patient-controlled anesthesia.

Table 6. Postoperative pain management after BABA RoT.

decrease thereafter; there was no significant difference ($P = 0.308$) between the 91 BABA RoT cases and 27 lobectomy cases. Moreover, Chai et al. reported that patient satisfaction in terms of voice quality, as assessed using an iPAD, was similar on days 1, 2, 3, and 14 between the two groups [49]. They also reported that the mean VHI10 scores on day 14 were similar between the BABA RoT and OT groups ($P = 0.849$). However, more large-scale prospective studies are needed to assess the difference in voice dysfunction between BABA RoT and OT. In addition, no studies have reported data pertaining to swallowing disorders after BABA RoT.

5.3.3. Sensory changes

There is concern regarding potential changes in the sensation of skin flaps caused by the more extensive dissection with BABA techniques compared with OT. In a prospective study by Kim et al., 19 patients underwent skin flap sensory assessments preoperatively and at 1 and 3 months postoperatively [52]. After BABA thyroidectomy, anterior chest paresthesia was normalized completely by 3 months. These results suggest that BABA has minimal adverse effects on anterior chest sensation.

5.4. Cosmetic satisfaction

The cosmetic outcome of the BABA technique involves practically no scarring, because this method transfers the anterior neck scar to four small hidden areas (the bilateral axilla and breasts), leaving the neck free of scars (**Figure 26**) [4]. Despite the early phase, we have already reported cosmetic satisfaction with endoscopic BABA according to a simple questionnaire [2]. Using an in-depth survey performed by a psychology consultant to evaluate neck scarring and psychological distress in patients who underwent BABA RoT, Koo et al. reported that the degree of scarring was significantly lower in the RoT group than the OT group ($P < 0.001$) [50]. There was also a significant difference in psychological distress between the immediate postoperative ($P = 0.009$) and follow-up period ($P < 0.001$). These results show the importance of scarless neck surgery and the esthetic superiority of the BABA method. In addition, Chai et al. reported significantly higher wound satisfaction scores in the 27 BABA RoT cases compared with the OT cases (7.4 vs. 5.7; $P = 0.016$) [49].

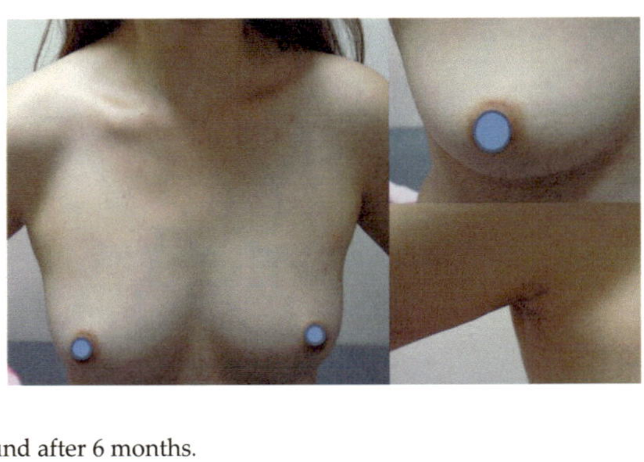

Figure 26. Postoperative wound after 6 months.

6. Special concerns for BABA RoT

6.1. BABA RoT experience with Graves' disease in comparison with OT

The application of ET for Graves' disease has been controversial. The major limitation is that it is not easy to control bleeding in cases of large hypervascular thyroid glands using non-flexible endoscopic instruments in a narrow two-dimensional field of view. Use of the surgical robot system has helped to overcome these limitations by introducing three-dimensional high-definition images and EndoWrist functions, which have resulted in more meticulous bleeding control. With recent technological advances and accumulation of experience, Kwon et al. reported successful results with comparable complication rates in 30 patients with Graves' disease [7]. There were no major complications, such as bleeding, open conversion, or permanent RLN injury, except for one case of permanent hypoPTH. In a subsequent article, Kwon et al. compared the safety of BABA RoT with that of OT in patients with Graves' disease ($n = 44$ and $n = 145$, respectively) and found comparable surgical completeness and complications between BABA RoT and conventional OT [53]. Therefore, BABA RoT may be a good surgical alternative for patients with Graves' disease who are concerned about cosmesis.

6.2. Influence of obesity on BABA RoT

Obesity is associated with various medical comorbidities that pose technical and clinical challenges, especially during surgery. For example, since a high BMI is a risk factor for various surgical complications, a retrospective study analyzed the influence of obesity on the surgical outcome of BABA RoT ($n = 310$) [10]. There was no statistically significant difference in body habitus indices, the length of hospital stay, surgical completeness, or complication rates between patients with a BMI <25 kg/m^2 and those with a BMI ≥ 25 kg/m^2. In conclusion, patients with a high BMI undergoing BABA RoT are not at increased risk of surgical complications; therefore, BABA RoT may be a good alternative for obese patients concerned with cosmesis.

Author details

Do Hoon Koo[1]*, Dong Sik Bae[1] and June Young Choi[2]

*Address all correspondence to: dohoon.koo.md@gmail.com

1 Department of Surgery, Inje University College of Medicine & Haeundae Paik Hospital, Busan, Korea

2 Department of Surgery, Seoul National University Bundang Hospital and College of Medicine, Seoul, Korea

References

[1] Shimazu K, Shiba E, Tamaki Y, et al. Endoscopic thyroid surgery through the axillo-bilateral-breast approach. Surgical Laparoscopy Endoscopy & Percutaneous Techniques. 2003;**13**:196-201

[2] Choe JH, Kim SW, Chung KW, et al. Endoscopic thyroidectomy using a new bilateral axillo-breast approach. World Journal of Surgery. 2007;**31**:601-606

[3] Chung YS, Choe JH, Kang KH, et al. Endoscopic thyroidectomy for thyroid malignancies: Comparison with conventional open thyroidectomy. World Journal of Surgery. 2007;**31**:2302-2306

[4] Lee KE, Rao J, Youn YK. Endoscopic thyroidectomy with the da Vinci robot system using the bilateral axillary breast approach (BABA) technique: Our initial experience. Surgical Laparoscopy, Endoscopy & Percutaneous Techniques. 2009;**19**:e71-75

[5] Chai YJ, Suh H, Woo JW, et al. Surgical safety and oncological completeness of robotic thyroidectomy for thyroid carcinoma larger than 2 cm. Surgical Endoscopy. 2017 Mar; **31**(3):1235-1240

[6] Kim BS, Kang KH, Kang H, et al. Central neck dissection using a bilateral axillo-breast approach for robotic thyroidectomy: Comparison with conventional open procedure after propensity score matching. Surgical Laparoscopy, Endoscopy & Percutaneous Techniques. 2014;**24**:67-72

[7] Kwon H, Koo doH, Choi JY, et al. Bilateral axillo-breast approach robotic thyroidectomy for Graves' disease: An initial experience in a single institute. World Journal of Surgery. 2013;**37**:1576-1581

[8] Lee KE, Kim E, Koo do H, et al. Robotic thyroidectomy by bilateral axillo-breast approach: Review of 1,026 cases and surgical completeness. Surgical Endoscopy. 2013; **27**:2955-2962

[9] Choi JY, Lee KE, Koo do H, et al. Use of spiral computed tomography volumetry for determining the operative approach in patients with Graves' disease. World Journal of Surgery. 2014;**38**:639-644

[10] Lee HS, Chai YJ, Kim SJ, et al. Influence of body habitus on the surgical outcomes of bilateral axillo-breast approach robotic thyroidectomy in papillary thyroid carcinoma patients. Annals of surgical treatment and research. 2016;**91**:1-7

[11] Bae DS, Koo doH, Choi JY, et al. Current status of robotic thyroid surgery in South Korea: A web-based survey. World Journal of Surgery. 2014;**38**:2632-2639

[12] Lee KE, Koo doH, Kim SJ, et al. Outcomes of 109 patients with papillary thyroid carcinoma who underwent robotic total thyroidectomy with central node dissection via the bilateral axillo-breast approach. Surgery. 2010;**148**:1207-1213

[13] Seup Kim B, Kang KH, Park SJ. Robotic modified radical neck dissection by bilateral axillary breast approach for papillary thyroid carcinoma with lateral neck metastasis. Head Neck. 2015;**37**:37-45

[14] Kang KH, Kim BS, Kang H. The benefits of preincision ropivacaine infiltration for reducing postoperative pain after robotic bilateral axillo-breast approach thyroidectomy: A prospective, randomized, double-blind, placebo-controlled study. Annals of Surgical Treatment and Research. 2015;**88**:193-199

[15] Kim HY, Choi YJ, Yu HN, et al. Optimal carbon dioxide insufflation pressure during robot-assisted thyroidectomy in patients with various benign and malignant thyroid diseases. World Journal of Surgical Oncology. 2012;**10**:202

[16] Jung SP, Kim SH, Bae SY, et al. A new subfascial approach in open thyroidectomy: Efficacy for postoperative voice, sensory, and swallowing symptoms. A randomized controlled study. Annals of Surgical Oncology. 2013;**20**:3869-3876

[17] Paek SH, Choi JY, Lee KE, et al. Bilateral axillo-breast approach (BABA) endoscopic Sistrunk operation in patients with thyroglossal duct cyst: Technical report of the novel endoscopic Sistrunk operation. Surgical Laparoscopy, Endoscopy & Percutaneous Techniques. 2014;**24**:e95-98

[18] Chai YJ, Kim SJ, Choi JY, et al. Papillary thyroid carcinoma located in the isthmus or upper third is associated with Delphian lymph node metastasis. World Journal of Surgery. 2014;**38**:1306-1311

[19] Lee HY, Lee JY, Dionigi G, et al. The efficacy of intraoperative neuromonitoring during robotic thyroidectomy: A prospective, randomized case-control evaluation. Journal of Laparoendoscopic & Advanced Surgical Techniques Part A. 2015;**25**:908-914

[20] Bae DS, Kim SJ. Intraoperative neuromonitoring of the recurrent laryngeal nerve in robotic thyroid surgery. Surgical Laparoscopy, Endoscopy & Percutaneous Techniques. 2015;**25**:23-26

[21] Yu HW, Chung JW, Yi JW, et al. Intraoperative localization of the parathyroid glands with indocyanine green and Firefly(R) technology during BABA robotic thyroidectomy. Surgical Endoscopy. 2016 Nov [Epub ahead of print].

[22] Kim SJ, Lee KE, Oh BM, et al. Intraoperative neuromonitoring of the external branch of the superior laryngeal nerve during robotic thyroid surgery: A preliminary prospective study. Annals of Surgical Treatment and Research .2015;**89**:233-239

[23] Kim WW, Jung JH, Park HY. A single surgeon's experience and surgical outcomes of 300 robotic thyroid surgeries using a bilateral axillo-breast approach. Journal of Surgical Oncology. 2015;**111**:135-140

[24] Bae DS, Kim SJ, Koo do H, et al. Prospective, randomized controlled trial on use of ropivacaine after robotic thyroid surgery: Effects on postoperative pain. Head Neck. 2016;**38**(Suppl 1):E588-593

[25] Kim WW, Kim JS, Hur SM, et al. Is robotic surgery superior to endoscopic and open surgeries in thyroid cancer? World Journal of Surgery. 2011;**35**:779-784

[26] Lee HY, Yang IS, Hwang SB, et al. Robotic thyroid surgery for papillary thyroid carcinoma: Lessons learned from 100 consecutive surgeries. Surgical Laparoscopy, Endoscopy & Percutaneous Techniques . 2015;**25**:27-32

[27] Cho JN, Park WS, Min SY, et al. Surgical outcomes of robotic thyroidectomy vs. conventional open thyroidectomy for papillary thyroid carcinoma. World Journal of Surgical Oncology. 2016;**14**:181

[28] Bae DS, Suh BJ, Park JK, et al. Technical, oncological, and functional safety of bilateral axillo-breast approach (BABA) robotic total thyroidectomy. Surgical Laparoscopy, Endoscopy & Percutaneous Techniques. 2016;**26**:253-258

[29] Kwak HY, Kim HY, Lee HY, et al. Robotic thyroidectomy using bilateral axillo-breast approach: Comparison of surgical results with open conventional thyroidectomy. Journal of Surgical Oncology. 2015;**111**:141-145

[30] Kim SK, Woo JW, Park I, et al. Propensity score-matched analysis of robotic versus endoscopic bilateral axillo-breast approach (BABA) thyroidectomy in papillary thyroid carcinoma. Langenbeck's Archives of Surgery. 2017 Mar;**402**(2):243-250

[31] Kandil E, Hammad AY, Walvekar RR, et al. Robotic thyroidectomy versus nonrobotic approaches: A meta-analysis examining surgical outcomes. Surgical Innovation .2016;**23**:317-325

[32] Lang BH, Wong CK, Tsang JS, et al. A systematic review and meta-analysis comparing surgically-related complications between robotic-assisted thyroidectomy and conventional open thyroidectomy. Annals of Surgical Oncology. 2014;**21**:850-861

[33] Son SK, Kim JH, Bae JS, et al. Surgical safety and oncologic effectiveness in robotic versus conventional open thyroidectomy in thyroid cancer: A systematic review and meta-analysis. Annals of Surgical Oncology. 2015;**22**:3022-3032

[34] Sun GH, Peress L, Pynnonen MA. Systematic review and meta-analysis of robotic vs. conventional thyroidectomy approaches for thyroid disease. Otolaryngology—Head and Neck Surgery: Official Journal of American Academy of Otolaryngology-Head and Neck Surgery. 2014;**150**:520-532

[35] Wang YC, Liu K, Xiong JJ, et al. Robotic thyroidectomy versus conventional open thyroidectomy for differentiated thyroid cancer: Meta-analysis. Journal of Laryngology and Otology. 2015;**129**:558-567

[36] Lee J, Yun JH, Choi UJ, et al. Robotic versus endoscopic thyroidectomy for thyroid cancers: A multi-institutional analysis of early postoperative outcomes and surgical learning curves. Journal of Oncology. 2012;**2012**:734541

[37] Hughes DT, White ML, Miller BS, et al. Influence of prophylactic central lymph node dissection on postoperative thyroglobulin levels and radioiodine treatment in papillary thyroid cancer. Surgery. 2010;**148**:1100-1106

[38] Randolph GW, Duh QY, Heller KS, et al. The prognostic significance of nodal metastases from papillary thyroid carcinoma can be stratified based on the size and number of metastatic lymph nodes, as well as the presence of extranodal extension. Thyroid: Official Journal of the American Thyroid Association. 2012;**22**:1144-1152

[39] Lee KE, Koo do H, Im HJ, et al. Surgical completeness of bilateral axillo-breast approach robotic thyroidectomy: Comparison with conventional open thyroidectomy after propensity score matching. Surgery. 2011;**150**:1266-1274

[40] Schlumberger M, Catargi B, Borget I, et al. Strategies of radioiodine ablation in patients with low-risk thyroid cancer. The New England Journal of Medicine. 2012;**366**:1663-1673

[41] Mallick U, Harmer C, Yap B, et al. Ablation with low-dose radioiodine and thyrotropin alfa in thyroid cancer. The New England journal of medicine. 2012;**366**:1674-1685

[42] Lombardi CP, Raffaelli M, de Crea C, et al. Report on 8 years of experience with video-assisted thyroidectomy for papillary thyroid carcinoma. Surgery. 2007;**142**:944-951; discussion 44-51

[43] Choi JY, Lee KE, Chung KW, et al. Endoscopic thyroidectomy via bilateral axillo-breast approach (BABA): Review of 512 cases in a single institute. Surgical Endoscopy. 2012;**26**:948-955

[44] Tae K, Song CM, Ji YB, et al. Comparison of surgical completeness between robotic total thyroidectomy versus open thyroidectomy. Laryngoscope. 2014;**124**:1042-1047

[45] Lee S, Lee CR, Lee SC, et al. Surgical completeness of robotic thyroidectomy: A prospective comparison with conventional open thyroidectomy in papillary thyroid carcinoma patients. Surgical Endoscopy. 2014;**28**:1068-1075

[46] Kim TY, Kim WB, Kim ES, et al. Serum thyroglobulin levels at the time of 131I remnant ablation just after thyroidectomy are useful for early prediction of clinical recurrence in low-risk patients with differentiated thyroid carcinoma. The Journal of Clinical Endocrinology and Metabolism. 2005;**90**:1440-1445

[47] Mazzaferri EL, Robbins RJ, Spencer CA, et al. A consensus report of the role of serum thyroglobulin as a monitoring method for low-risk patients with papillary thyroid carcinoma. The Journal of Clinical Endocrinology and Metabolism. 2003;**88**:1433-1441

[48] Lang BH, Wong CK, Tsang JS, et al. A systematic review and meta-analysis evaluating completeness and outcomes of robotic thyroidectomy. Laryngoscope. 2015;**125**:509-518

[49] Chai YJ, Song J, Kang J, et al. A comparative study of postoperative pain for open thyroidectomy versus bilateral axillo-breast approach robotic thyroidectomy using a self-reporting application for iPad. Annals of Surgical Treatment and Research. 2016;**90**:239-245

[50] Koo do H, Kim da M, Choi JY, et al. In-depth survey of scarring and distress in patients undergoing bilateral axillo-breast approach robotic thyroidectomy or conventional open thyroidectomy. Surgical Laparoscopy, Endoscopy & Percutaneous.Techniques. 2015;**25**:436-439

[51] Ryu JH, Yom CK, Kwon H, et al. A prospective, randomized, controlled trial of the postoperative analgesic effects of spraying 0.25 % levobupivacaine after bilateral axillo-breast approach robotic thyroidectomy. Surgical Endoscopy. 2015;**29**:163-169

[52] Kim SJ, Lee KE, Myong JP, et al. Prospective study of sensation in anterior chest areas before and after a bilateral axillo-breast approach for endoscopic/robotic thyroid surgery. World Journal of Surgery. 2013;**37**:1147-1153

[53] Kwon H, Yi JW, Song RY, et al. Comparison of bilateral axillo-breast approach robotic thyroidectomy with open thyroidectomy for graves' disease. World Journal of Surgery. 2016;**40**:498-504

Chapter 5

Colonoscopy Image Pre-Processing for the Development of Computer-Aided Diagnostic Tools

Alain Sánchez-González and
Begoña García-Zapirain Soto

Additional information is available at the end of the chapter

http://dx.doi.org/10.5772/67842

Abstract

Colorrectal cancer is the third most frequently diagnosed cancer worldwide. The American Cancer Society estimates that there will be almost 100,000 new patients diagnosed with colorectal cancer and that around 50,000 people will die as a consequence of this in 2016. The increase of life expectancy and the increment of the number of diagnostic tests conducted have had a great impact on the amount of cancers being detected. Among other diagnostic tools, colonoscopy is the most prevalent. In order to help endoscopists cope with the increasing amount of tests that have to be carried out, there exists a need to develop automated tools that aid diagnosis. The characteristics of the colon make preprocessing essential to eliminate artefacts that degrade the quality of exploratory images. The goal of this chapter is to describe the most common issues of colonoscopic imagery as well the existing methods for their optimal detection and correction.

Keywords: colonoscopy, medical image pre-processing, specular reflections, inhomogeneous illumination, black borders, interlacing

1. Introduction

The unceasing increase in incidences of colorectal cancer (CRC) in recent decades has led to a rise in the number of medical tests being carried out; in the case at hand, colonoscopies. Specialists consequently have a greater amount of work, and find themselves overwhelmed. As a result of this problem, numerous investigations have been conducted in recent years focussing on developing tools to help with diagnoses, thereby supporting specialists. Development of algorithms for the automatic analysis of colonoscopy imaging requires preliminary preprocessing of the images in order to rectify the multiple factors that detract from their quality.

The objective of this chapter is to shed light on the most common problems encountered in colonoscopic imaging, while also providing the most frequentlys-used solutions among the scientific community. The aim is to thusly supply useful information in order to develop automatic algorithms, which may then be implanted in robots that automate tasks currently requiring manual interaction.

2. What is a colonoscopy?

A colonoscopy is a method of reference for diagnosing and treating colonic diseases; essential to both colorectal screening and monitoring. This exploration enables the large intestine to be viewed in its entirety, to extract biopsies and to remove tumours.

It has been proven that carrying out this procedure reduces the colon cancer mortality rate. Before undergoing the procedure, it is necessary for the patient to have been through a preparation phase, so that there is no solid waste in the colon. The procedure is performed by inserting a colonoscope—a flexible tube with a camera at the end—into the anus (see **Figure 1**). In some cases, a sedative is used so as to carry out the procedure without causing discomfort. It is the best means of detecting CRC since it enables localisation and, in the majority of cases, immediate extraction.

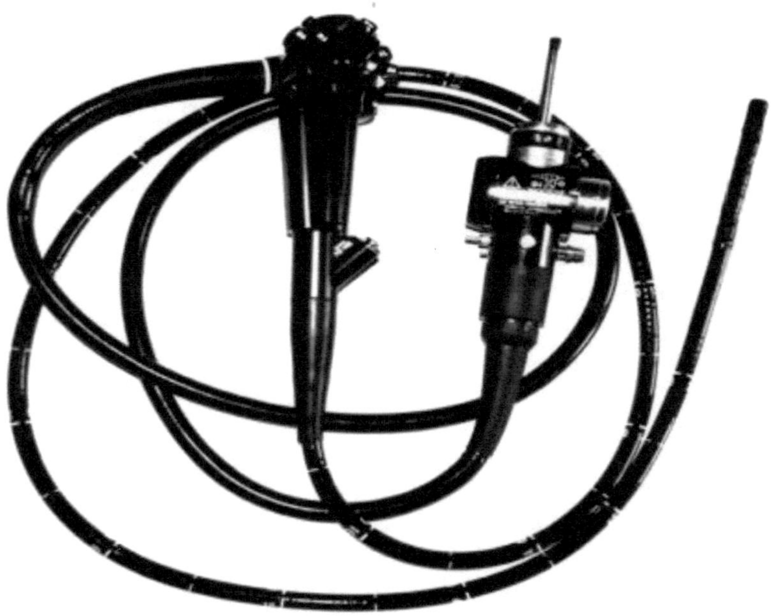

Figure 1. Colonoscope. *Source*: goo.gl/6qtSW9.

3. Main problems of colonoscopies

The principal difficulties in obtaining colonoscopy imaging are described below; which, in many cases, are the result of the equipment used or the environmental difficulties.

Black mask: this is due to the fact that the lenses used in the colonoscopy image capturing system have a black frame around the edge. In many cases, the mask is used to convey information, either pertaining to the patient or the test being carried out. This black frame hinders the development of digital image processing algorithms since it creates false borders, as well as covering a larger area for analysis that would not yield useful information. For these reasons, applying different techniques to eliminate its effects becomes necessary. In **Figure 2**, the black mask in colonoscopy imaging can be observed.

Ghost colours: the problem of ghost colours (see **Figure 3**) is linked to a lack of synchronisation of the colour channels. Its appearance is due to the fact that most colonoscopy equipment uses monochromatic cameras, in which the components R, G, and B are obtained at different times. This causes a reduction in the quality of the image, making the subsequent development of PDI algorithms difficult.

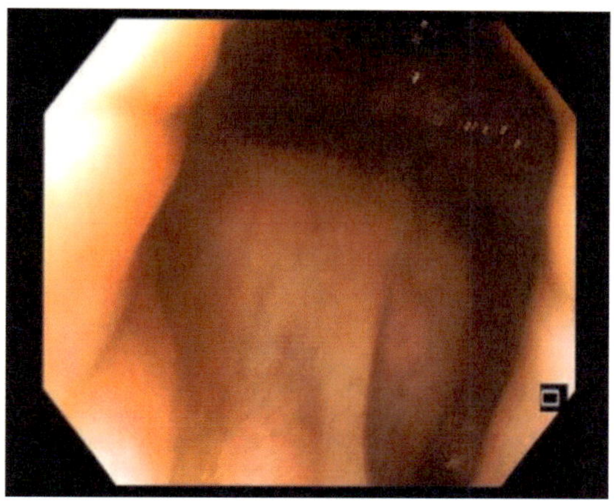

Figure 2. Black mask.

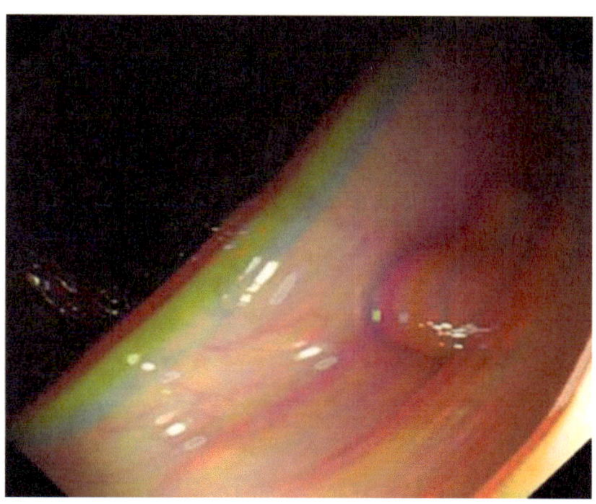

Figure 3. Ghost colours.

Interlacing: interlacing allows twice the number of frames per second to be taken without consuming additional bandwidth. It is used in standard formats such as the National Television System Committee (NTSC) or phase alternating line (PAL), and shows half of the horizontal lines in each iteration. Each frame is divided into two fields: the first contains odd-numbered lines and the second field the even-numbered lines. Due to the phenomenon of the persistence of the human eye, the brain mixes both iterations of the interlaced frame, identifying it as one image. The effects of interlacing cause the appearance of false outlines in the images (see **Figure 4**), which make the development of algorithms more complicated. Therefore, it is necessary to implement techniques to reduce its occurrence.

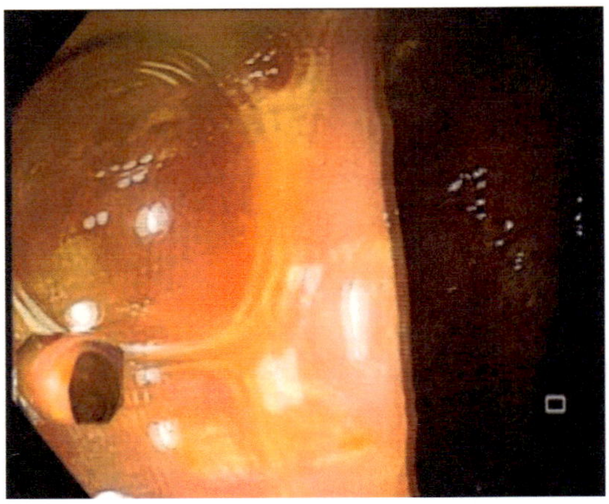

Figure 4. Effects of interlacing.

Specular highlights: specular highlights (see **Figure 5**) are points of high intensity in the image due to the illumination of shiny objects. When a source of light is shone directly on an object, the light is reflected and captured by the camera. This process generates heavily saturated

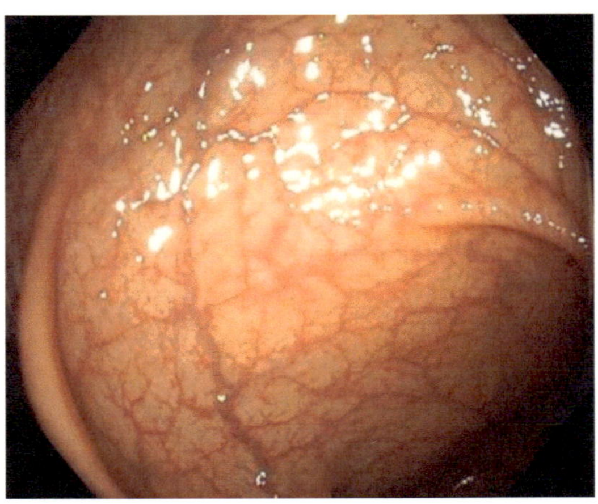

Figure 5. Specular highlights.

http://dx.doi.org/10.5772/67842

areas in the image, which can lead to unwanted outlines, making it subsequently difficult to process the image. This effect is extremely important in the detection of polyps, which are generally rounded and similar to tumours. Due to their shape, they reflect light and generate specular highlights when illuminated, which can lead to a malfunction of the developed algorithms.

Uneven lighting: the variations in the intensity and direction of lighting are decisive in the appearance of objects in digital images. The illumination of the colon in a colonoscopy is variable, which, because of the colon's three-dimensional shape, causes shadows to appear, accentuating or diminishing certain aspects of the image. Varying degrees of illumination on the same object cause differing representations of the object, rendering said variability of lighting unwanted. In the literature, there are numerous publications that address this problem. In **Figure 6**, an example of uneven lighting in colonoscopic imaging is shown in order to facilitate its detection.

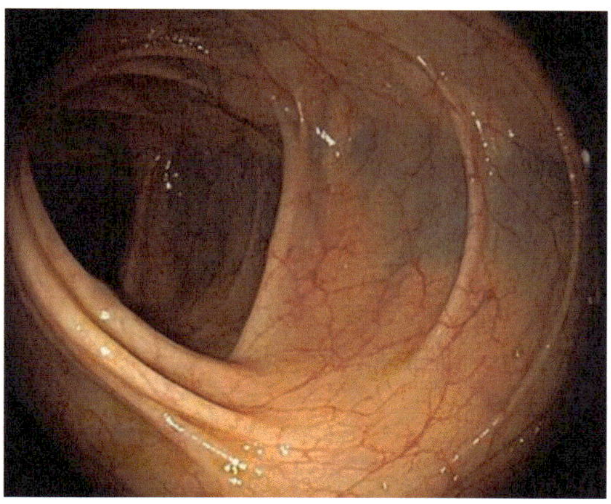

Figure 6. Uneven lighting.

4. What is the pre-processing of colonoscopic images?

Every image capturing process is affected in some way by factors that reduce the quality of the image to some degree. Colonoscopic imaging is no exception, so it is necessary to implement techniques that help to improve the quality and thereby obtain a better visual representation.

Any technique whose objective is to contrast, highlight, accentuate or remove unwanted effects from the image is considered a method of improvement. This is a process of vital importance in medical imaging, in which the limitations of the image capturing system—in the case at hand, colonoscopies—cause unwanted effects which need to be removed. It is crucial to point out that by improving imaging:

- No new information is added to the image; the image is only highlighted so as to be used more efficiently by the algorithms that are to be developed.

- There is no exact criterion for quantifying the degree of improvement; in many cases, it is based on subjective opinions and/or on the result obtained by the developed algorithms.

Below is an outline of the applicability of pre-processing colonoscopic imagery in robots which may be able to automate tasks that are vital in a colonoscopy.

5. Applicability of pre-processing colonoscopic imagery in robots

Faced with the growing number of diagnostic tests for colon cancer being carried out, it has become necessary to rely on support tools for medical diagnoses. These tools support the specialist by providing objective data, thereby enabling more accurate diagnoses.

The main functions that endoscopists require are related to the automatic detection of polyps and the evaluation of the quality of the test being carried out.

In the case of detecting polyps, having tools available that enable their automatic detection will mean a reduction in the number of missed tumours, which, in many cases, lead to interval cancers. Interval cancers are those that appear between two scheduled diagnostic tests and, in most cases, are due to a polyp or tumour that was not detected by the specialist during the procedure. In this context, publications such as [1–3] have made important contributions to the scientific community.

Moreover, the quality evaluation of the procedure is a necessity, since many of the metrics are currently based on the specialist's interpretation and are therefore subjective, impeding correct comparison among different health centres with the intention of improving the process. The European Guidelines for Quality Assurance in Colorectal Screening and Diagnosis [4] provide a series of metrics that evaluate different aspects of the colonoscopy. In this regard, publications such as [5–8] make valuable contributions to the scientific community.

All research studies focused on the development of automated tools for the assistance of medical diagnoses share the need for the availability of an image pre-processing system. The availability of tools to improve the quality of the images is a necessity, as can be observed in investigation [9].

All the methods for pre-processing imagery outlined in this chapter will be able to be implanted in robots and colonoscopies in such a way as to enable the development of various automated tools, which allows for significant higher reliability of colonoscopies.

6. Pre-processing colonoscopic imagery

Here, we describe the most frequently used techniques in the scientific community for removing the most common discrepancies in colonoscopic imagery. Solutions that have been proposed in the literature are outlined, and the most appropriate focus for each point has also been proposed.

http://dx.doi.org/10.5772/67842

6.1. Removal of black borders

In the literature, there are three tendencies for black border removal: the restoration of the image, the use of thresholding and cropping of the black mask. Following is a brief explanation of each method.

- Removal of the black mask through restoring the image: this involves replacing the pixels of the black mask using the median value of the pixels in a certain vicinity. This focus has been used in investigation [9], obtaining satisfactory results.
- Removal of the black mask using thresholding: a threshold is set to detect the real frame of the image, removing the black mask. In many occasions, this focus does not manage to completely remove the black mask, leaving residual lines, which makes it necessary to apply techniques such as the Hough transform [10] to remove them. This technique was used in investigation [11].
- Removal of the black mask through cropping the image: this is the simplest focus, in which an area of the image is selected and the rest is removed. This method involves obtaining a smaller image but maintaining the maximum amount of information possible from the original image, running the risk of losing valuable information.

In this section, a suggestion for an alternative focus for the removal of black borders is presented. Depending on the model of colonoscope used, the black borders that are generated vary (see **Figure 7**), which makes pre-processing difficult. In many cases, the borders are used to provide information about either the patient or the procedure being carried out (see **Figure 7(b)**). This frame makes the development of PDI algorithms difficult, since it generates false borders, as well as entailing a greater area to be analysed that does not provide useful information. Due to these reasons, it is necessary to apply different techniques to remove their effects.

There are various literary references to methods addressing this problem: reconstructing the borders by restoring them [9], the use of thresholding for their detection [11] and the cropping of the black mask. In this pre-processing design, a method combining the existing solutions was chosen. This technique involves detecting the black mask using thresholding, as well as cropping and reconstructing. **Figure 8** shows the process in which this task is carried out.

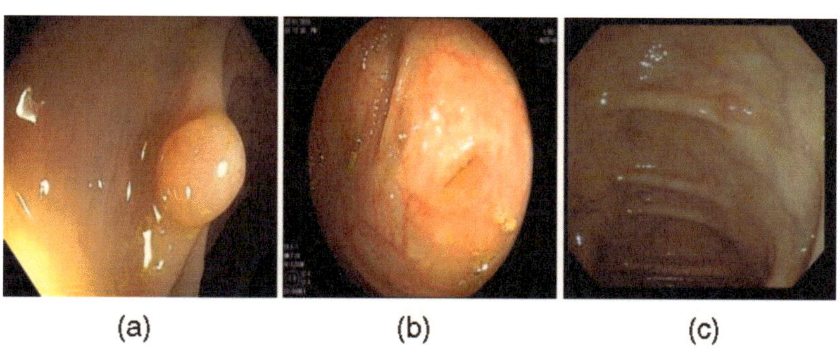

Figure 7. Black masks with different characteristics: (a) Black mask in the corners. (b) Circular black mask and with information. (c) Black mask bordering the image.

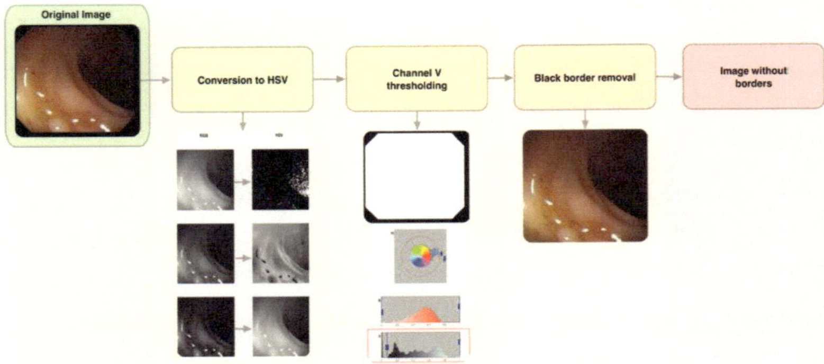

Figure 8. Removal of black borders.

The following is a description of the steps to remove the black borders using the proposed method:

(1) *Conversion to Hue, Saturation, Value:* in order to address the automatic detection of the black mask in colonoscopic images, it is necessary to convert them from the RGB colour model (the original colour model for colonoscopic imagery) to the HSV colour model. This is due to the fact that the RGB model makes certain colour specification difficult, whereas this is one of the HSV model's strengths. Thanks to this, the thresholding described in the next step is made much more simple.

(2) *Channel V thresholding*: once the conversion from the RGB colour model to HSV is complete, the image is ready for thresholding. Thresholding offers a wide range of intensity values from which to choose, allowing us to define among them those objects that we want to be detected automatically. In this chapter, channel V thresholding is proposed, in which values of 0.03 and lower are attributed to the black mask. This method enables the separation of useful content in the colonoscopic image from the black borders. This process can be observed in **Figure 9**, in which **Figure 9(a)** shows the process of Channel V thresholding and **Figure 9(b)** presents the result generated.

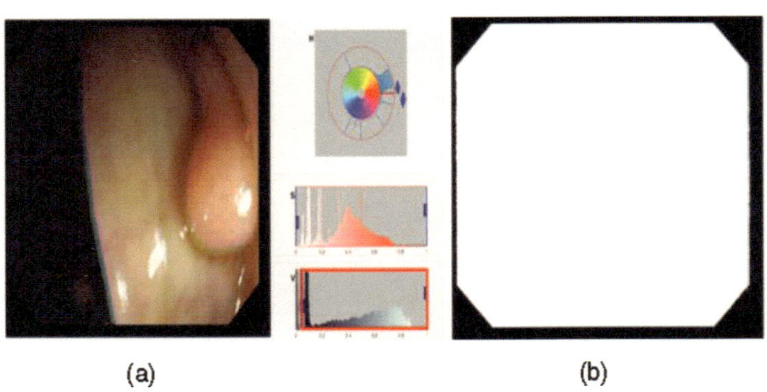

Figure 9. Thresholding for the detection of black borders: (a) Channel V thresholding highlighted in red. (b) Result of thresholding.

Depending on the model of colonoscope used to capture the images, the black borders may be different. This is a problem, since when thresholding is carried out to detect the black borders, the information shown in the borders will remain visible over the image. In order to remove it, an additional step is necessary which involves making a morphological opening by using a size-5 disk structure to the detected black mask. In this way, all the information shown on the black border is removed, leaving it clean. This process can be observed in **Figure 10**, in which **Figure 10(a)** shows the detected black mask with leftover information and **Figure 10(b)** shows the result of the morphological opening for its removal.

Once the thresholding of the image is complete, it is possible to proceed to the removal of the black borders.

(3) *Black border removal:* the process of black border removal comprises two steps: cropping and reconstructing. The following is a detailed description of both:

- *Detection of the upper central point not belonging to the black mask*: starting from the pixel in position (max(X)/2.1) searching southwards, the first pixel does not belong to the black mask.
- *Detection of the lower central point not belonging to the black mask*: starting from the pixel in position (max(X)/2.max(Y)) searching northwards, the first pixel does not belong to the black mask.
- *Detection of the centre-left point not pertaining to the black mask*: starting from the pixel in position (1.max(Y)/2) searching eastwards, the first pixel does not belong to the black mask.
- *Detection of the centre-right point not pertaining to the black mask*: starting from the pixel in position (max(X).max(Y)/2) searching westwards, the first pixel does not belong to the black mask.

Once the four positions of the sought pixels have been obtained, a rectangle is generated whi ch contains them and will be what determines the dimensions of the image with the black borders cropped out. **Figure 11** shows a visual example of this process. The next step in removing the black borders is the reconstruction of the leftover black borders. This process is addressed in the following section.

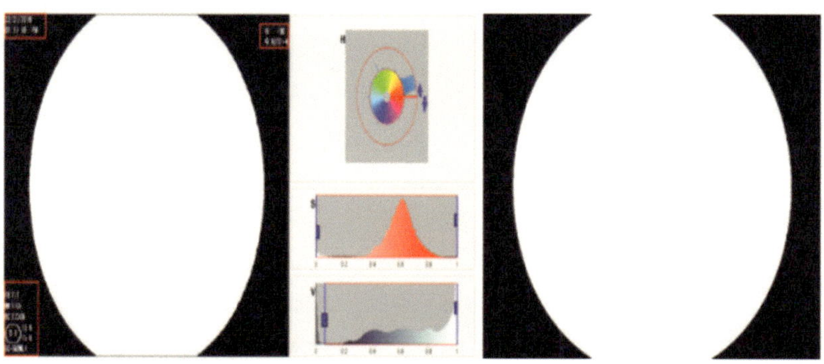

(a) (b)

Figure 10. Thresholding for the detection of black borders: (a) Black borders with information highlighted in red. (b) Result of the morphological opening.

Reconstruction of the remnants of the black mask: in **Figure 11** it can be seen that the final area of the image highlighted in orange still contains remnants of the black borders. The final task for their removal is to reconstruct them. In order to do so, a restoration is applied which aims to replace the pixels of the black mask by the median value of the pixels in a certain vicinity. This operation is carried out repeatedly until the difference between the values of the neighbouring pixels used in the reconstruction falls below a predetermined amount.

(4) *Image without black borders*: having performed all the procedures designed for black border removal, we will obtain an image with reduced dimensions and the reconstructed black borders. The result obtained can be seen in **Figure 12**, in which **Figure 12(a)** shows the original image without editing, and **Figure 12(b)** provides the result obtained through this process.

6.2. Removal of specular highlights

There are numerous methods to detect specular highlights. The following is a brief summary of the most important of these:

Park et al. [12] propose the detection of specular highlights using a search of saturated areas and small regions with high contrast. The saturated areas are detected by applying adaptive thresholding to the image's intensity histogram. The value of the threshold is predetermined

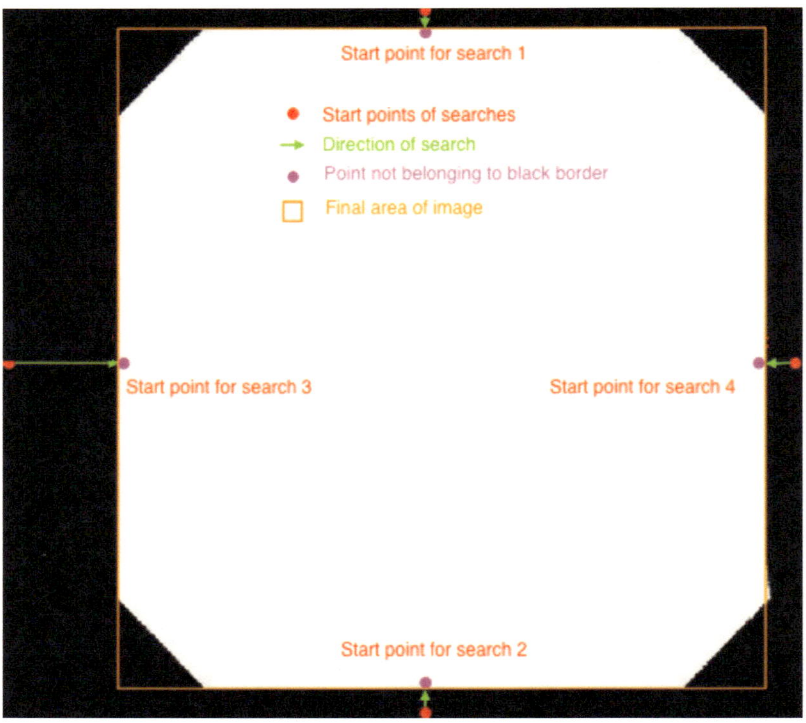

Figure 11. Process of cropping the black borders.

http://dx.doi.org/10.5772/67842

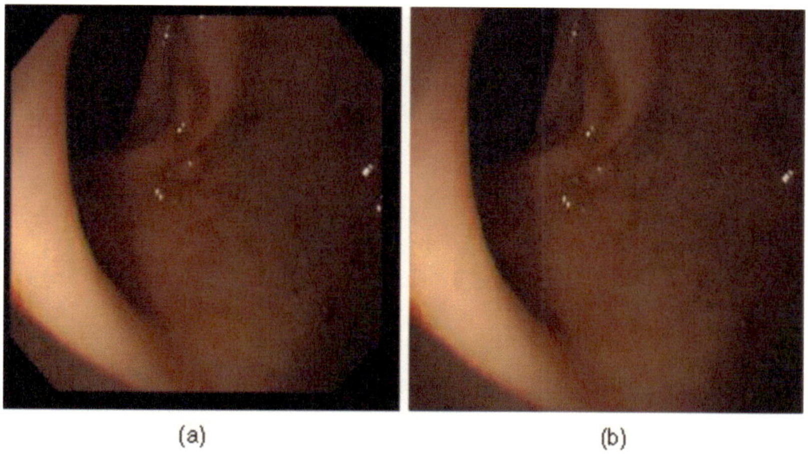

Figure 12. Result of black border removal: (a) Original image with black borders. (b) Result of black border removal.

as the region that surrounds the maximum value of the histogram. The smaller regions with high contrast are detected using the method proposed in Ref. [13], which applies a top-hat filter followed by a reconstruction and erosion operation by a size-5 disk structure.

Bernal et al. [9] assume that the specular pixel intensity value is greater than that of the non-specular pixels in their vicinity. Furthermore, they indicate that non-specular pixels which neighbour specular pixels will have higher intensity values than non-specular pixels far from the reflective areas. The detection of specular highlights is carried out by the subtraction of the original image and their median. Once this has been done, specular highlights can be detected through the use of thresholding.

Gross et al. [14] detect specular highlights based on the space of HSV colour. Specular highlights show a high saturation and low brightness, which makes their detection simple.

The method put forward in Ref. [15] for the detection of specular highlights uses two different colour spaces. In the first, it is necessary to observe the borders generated by the changes in texture and specular highlights. In the second, only the borders generated by the textures need to be seen. Subtraction of these two colour spaces enables the detection of specular highlights. This method has been used in investigation [16] with satisfactory results. Therein, the detection of specular highlights based on low saturation of the colour of the highlights is suggested.

Having shown the techniques used in various studies for the removal of specular highlights, the method for their elimination is proposed. **Figure 13** shows the steps for a better understanding. A description of each of the modules that comprise them follows.

(1) *Conversion to greyscale:* in order to commence the process of specular highlight removal, it is necessary to convert the borderless image from the original colour model (RGB) to greyscale. This operation is necessary for subsequent detection of specular highlights, which is described in the next step.

(2) *Detection of specular highlights:* the method used for the detection of specular highlights has been proposed by the authors of the study [9]. To this end, a system comprising four blocks has been designed, which is shown in **Figure 14**. In the following steps, there is a detailed description of the process for specular highlight removal proposed for this investigation.

- *Calculation of the threshold value (U):* to detect specular highlighting automatically, it is vital to affix a threshold value (U) which distinguishes between normal values in the image and specular highlighting. To this end, the median value of the original image (μ) is calculated on a greyscale, which is then multiplied by a weight (*W*) which, by default, has a value of 0.3. In this way, the value required for addressing the next phase in the detection of specular highlights is calculated.
- *Subtraction of the original image in greyscale and the threshold value:* once the threshold value (U) has been calculated, the subtraction of the original image in greyscale with the threshold value (U) is performed. In this way, a matrix equal in dimensions to that of the image in greyscale is obtained, in which values above 0.75 belong to specular highlighting.
- *Thresholding:* having calculated the matrix with the values pertaining to the subtraction between the original image in greyscale and the threshold value (U), a binary mask will be generated in which values surpassing the threshold (U) are given a value of 1, and everything else a value of 0, thereby obtaining an image that only shows the positions of the specular highlighting that has been detected.
- *Mask with specular highlights*: as a result of this process, a mask is obtained which will be used in the next step and will deal with the reconstruction of the highlighting.

(3) *Reconstruction of the image:* once the dilation of the specular highlighting mask has been carried out, we can begin to reconstruct the regions of the image indicated by the mask through the following steps:

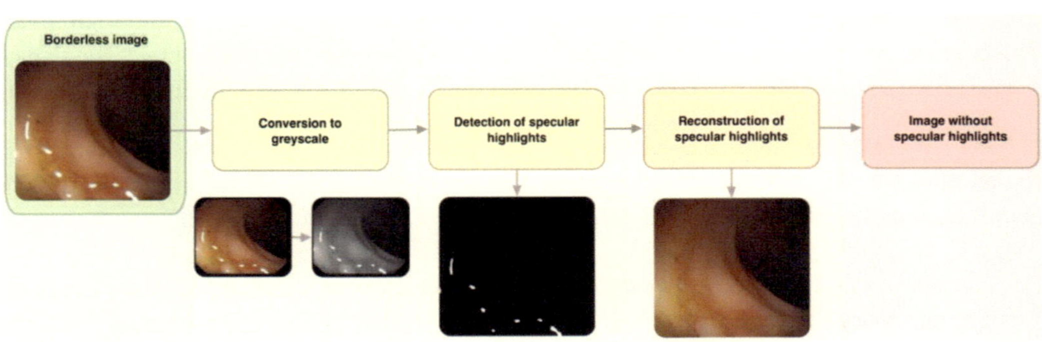

Figure 13. System for the removal of specular highlights.

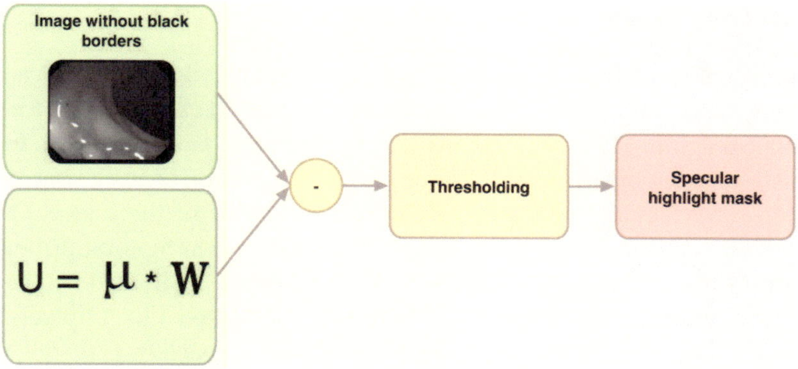

Figure 14. Detection of specular highlights. μ Represents the median value of the image without black borders and W denotes the multiplication factor (0.3 by default).

(1) The damaged section is filled in using information from the rest of the image.

(2) The structure of the area surrounding the deteriorated part is filled in towards its centre, extending the lines that reach the border.

(3) The numerous regions that are generated inside the damaged area from the extension of the contour lines are filled in with the colour of the corresponding bordering region.

(4) Finally, the small details are coloured in to maintain uniformity.

The algorithm repeatedly carries out steps 2 and 3 until the desired quality is achieved. Having carried out this process, an image free of specular highlights is obtained. The result is highly effective, as shown in **Figure 15**.

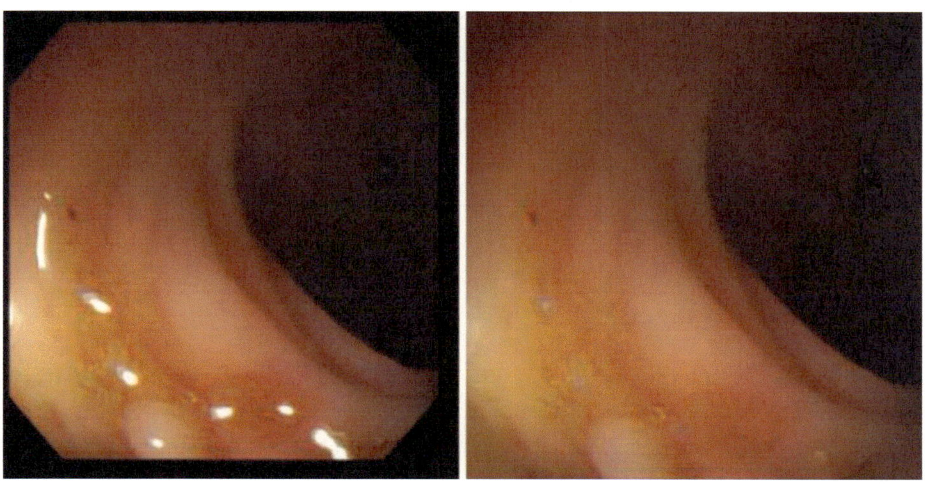

Figure 15. Removal of specular highlights.

6.3. Lighting normalisation

In the scientific literature, there are numerous publications that deal with uneven lighting in imaging. A brief summary of the most relevant works, as well as a proposal for an alternative to normalise lighting in colonoscopic imagery illumination is presented. Investigation [17] presents a contrast operator built by means of two primitives involving Weber's law, and, in doing so, achieving an improvement in the contrast of the image. On the other hand, study [16] carries out a reduction of the effects of uneven lighting through the local normalisation of the image's brightness. For this, each pixel is divided by the maximum value of its vicinity. In this publication, vicinity was considered 13 × 13 pixels. Finally, in investigation [14], an equalisation of the background of the image in greyscale was carried out, thereby strengthening the contrast of the different structures, as well as removing the lighting variation in the image.

The following procedure is proposed to solve the issue of homogenous lighting in colonoscopic imagery. The proposed design is shown in **Figure 16**, offering a complete description of the blocks comprising it; i.e. obtaining the subtraction value, subtracting the image with the subtraction value and the image with normalised lighting.

1. *Obtaining the subtraction value:* in order to achieve a more uniform illumination in the images, it is a fundamental requirement to calculate a subtraction value for each of the windows into which the image has been divided (20 × 20 pixels). This value is obtained by calculating the median value of each channel inside the said window and multiplying it by a weight (0.3 by default).

2. *Subtraction from the window with the subtraction value:* once the subtraction values of the different channels have been calculated, these are subtracted from the corresponding channel of the window. In this way, the effects of the peaks of intensity that the uneven lighting causes are mitigated.

3. *Image with normalised lighting:* as an output of the lighting normalisation module, an image is obtained with a range of much more uniform colour intensities, which aids its subsequent analysis. Following this previous step, the colonoscopic images are ready to be used for quality evaluation algorithms for the preparation of the colon, using the BBPS, and automatic polyp detection. In **Figure 17**, it is possible to observe the result obtained through the normalisation of lighting. **Figure 17(a)** shows an image without lighting normalisation and **Figure 17(b)** shows the result obtained through this process.

6.4. Removal of interlacing effects

The adverse effects of interlacing are habitual in the use of videos, or in the extraction of images from video frames. The removal of these aspects has been addressed in numerous investigations, which achieve very accurate results. Below, the most relevant publications that propose a solution to this problem are shown.

Studies [18–20] address the removal of the effects of interlacing through deinterlacing. The procedure is based on obtaining one in every two horizontal lines, decreasing the vertical

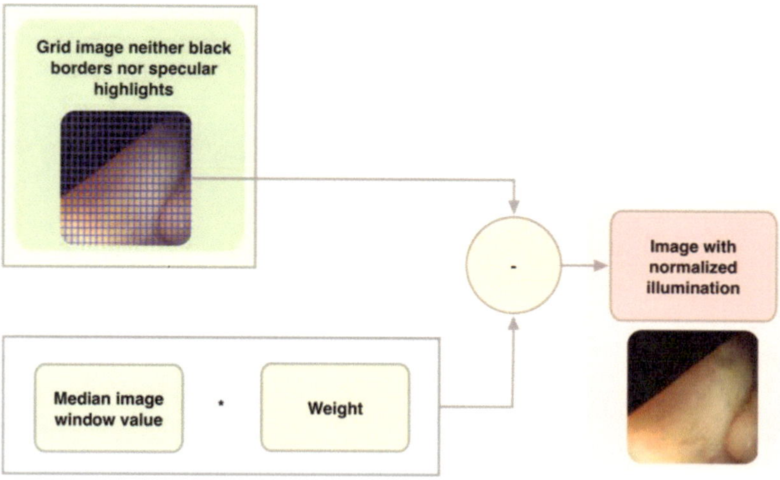

Figure 16. Illumination normalisation.

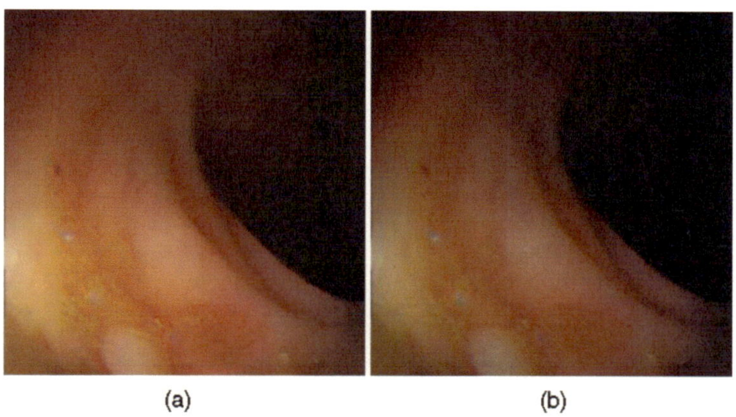

Figure 17. Result of lighting normalisation: (a) Image with neither black borders nor specular highlighting. (b) Result of lighting normalisation.

size of the image. To maintain the size proportion of the original image, they apply vertical redimensioning by a factor of 0.5.

Figure 18 shows the results of applying these techniques for removing interlacing effects. As can be observed, the obtained result is very good, achieving high effectiveness.

6.5. Removal of ghost colours

This problem has been addressed in the literature, in study [21], where channel equalisation is proposed, as is carrying out an estimation and compensation of the movements of the camera. Channel equalisation aims to obtain a histogram with a more uniform distribution,

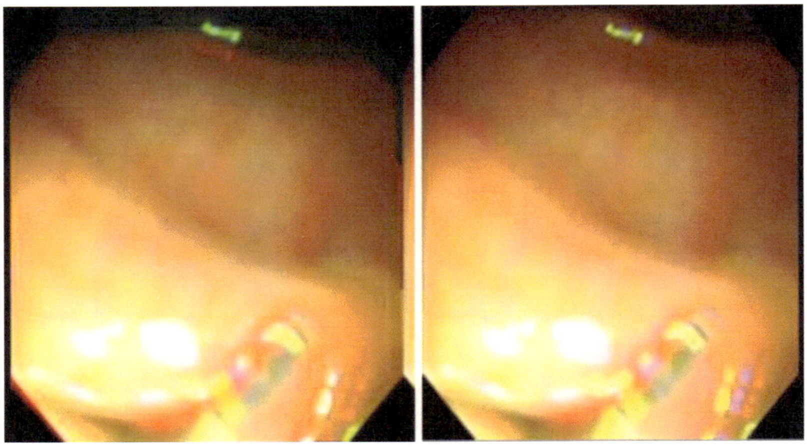

Figure 18. Removal of the effects of interlacing.

i.e. the same number of pixels should exist for each level of grey in the histogram of a monochrome image. The estimation and compensation of the movements of the camera are obtained through the use of the movement vectors from MPEG video standard. These enable an estimate of the deviation affecting each colour channel in obtaining the image, allowing the errors produced to be corrected. This same solution has been addressed in study [22]. The application of this technique corrects the effect very accurately, failing solely in images of very low initial quality. The result obtained using this solution is shown in **Figure 19**.

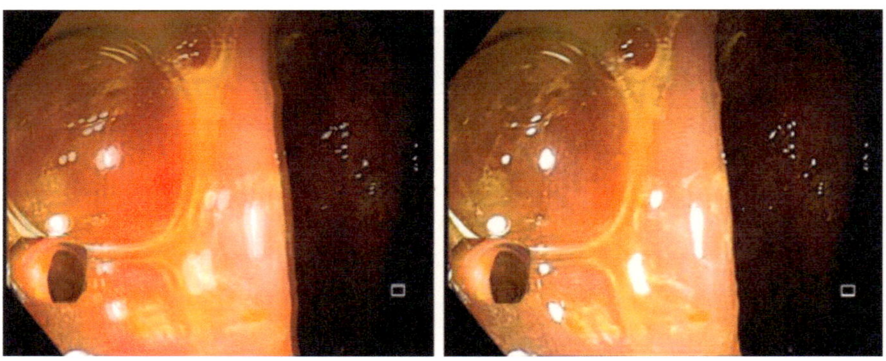

Figure 19. Removal of ghost colours.

7. Conclusion

The benefits derived from the tools described in the present chapter are in the improvement of colonoscopic images. Specifically:

- The scientific community is provided with information about the origin and characteristics of the most prevalent artefacts that corrupt colonoscopic images, thus allowing for their identification, detection and removal.
- The techniques that have to be applied to the images in order to increase their quality are described, as well as the methodology that has to be used to apply them.
- The scientific community is also given a useful guide to a system of medical diagnosis aid based on colonoscopic images, thus allowing to offer tools better suited to the needs of the patients.

Since the systems to aid diagnosis are constantly on the rise nowadays and are likely to be in the immediate future, we consider the current chapter is undoubtedly necessary to the specialist in the area.

Author details

Alain Sánchez-González* and Begoña García-Zapirain Soto

*Address all correspondence to: sanchez.alain@deusto.es

DeustoTech-LIFE Unit, University of Deusto, Biscay, Spain

References

[1] S.-H. Bae and K.-J. Yoon, Polyp detection via imbalanced learning and discriminative feature learning, IEEE Trans Med. Imaging., vol. 34, no. 11, pp. 2379-2393, 2015.

[2] J. Bernal, F. J. Sánchez, G. Fernández-Esparrach, D. Gil, C. Rodríguez, and F. Vilariño, WM-DOVA maps for accurate polyp highlighting in colonoscopy: Validation vs. saliency maps from physicians, Comput. Med. Imaging Graph, vol. 43, pp. 99-111, 2015.

[3] L. Alexandre, N. Nobre, J. Casteleiro, and others, Color and position versus texture features for endoscopic polyp detection, International Conference on BioMedical Engineering and Informatics, 2008. BMEI 2008, Sanya, China. vol. 2, pp. 38-42, 2008.

[4] S. P. Halloran, G. Launoy, and M. Zappa, European guidelines for quality assurance in colorectal cancer screening and diagnosis, Endoscopy, vol. 10, pp. 0032-1309791, 2012.

[5] N. F. Fayad and C. J. Kahi, colonoscopy quality assessment, Gastrointest. Endosc. Clin. N. Am., vol. 25, no. 2, pp. 373-386, 2015.

[6] J. Oh et al., Measuring objective quality of colonoscopy, IEEE Trans. Biomed. Eng., vol. 56, no. 9, pp. 2190-2196, 2009.

[7] M. F. Kaminski et al., Quality indicators for colonoscopy and the risk of interval cancer, N. Engl. J. Med., vol. 362, no. 19, pp. 1795-1803, 2010.

[8] S. Hwang, J. Oh, W. Tavanapong, J. Wong, and P. C. de Groen, Stool detection in colonoscopy videos, in Engineering in Medicine and Biology Society, 2008. EMBS 2008. 30th Annual International Conference of the IEEE, Vancouver Convention and Exhibition Centre,Vancouver, BC, Canada. pp. 3004-3007, 2008.

[9] J. Bernal, J. Sánchez, and F. Vilarino, Impact of image preprocessing methods on polyp localization in colonoscopy frames, in Engineering in Medicine and Biology Society (EMBC), 2013. 35th Annual International Conference of the IEEE, Osaka International Convention Center, in Osaka, Japan. pp. 7350-7354, 2013.

[10] R. O. Duda and P. E. Hart, Use of the Hough transformation to detect lines and curves in pictures, Commun. ACM, vol. 15, no. 1, pp. 11-15, 1972.

[11] H. C. Wang, W. M. Chen, Y. P. Lin, and W. C. Shen, Tumor detecting in colonoscopic narrow-band imaging data, presented at the 2012 International Symposium on Intelligent Signal Processing and Communications Systems (ISPACS), New Taipei City, Taiwan. pp. 564-568, 2012.

[12] S. Y. Park, D. Sargent, I. Spofford, K. G. Vosburgh, and Y. A-Rahim, A colon video analysis framework for polyp detection, IEEE Trans. Biomed. Eng., vol. 59, no. 5, pp. 1408-1418, 2012.

[13] D. S. Bright and E. B. Steel, Two-dimensional top hat filter for extracting spots and spheres from digital images, J. Microsc., vol. 146, no. 2, pp. 191-200, 1987.

[14] S. Gross, S. Palm, J. J. Tischendorf, A. Behrens, C. Trautwein, and T. Aach, Automated classi cation of colon polyps in endoscopic image data, in SPIE Medical Imaging, Orlando, Florida, United States. pp. 83150W-83150W, 2012.

[15] T. Gevers, Re ectance-based classi cation of color edges, in Computer Vision, 2003. Proceedings of Ninth IEEE International Conference, Washington, DC, USA. pp. 856-861, 2003.

[16] T. Stehle et al., Classi cation of colon polyps in NBI endoscopy using vascularization features, in SPIE Medical Imaging, Lake Buena Vista, Florida, United States. pp. 72602S-72602S, 2009.

[17] E. A. Rivas Araiza, J. D. Mendiola Santibañez, G. Herrera Ruiz, C. A. González Gutiérrez, M. Trejo Perea, and G. J. Ríos Moreno, Mejora de Contraste y Compensación en Cambios de la Iluminación, Comput. Sist., vol. 10, no. 4, 2007.

[18] Bernal, J., Sánchez, J., Vilariño, F.: Reduction of Pattern Search Area in Colonoscopy Images by Merging Non-Informative Regions. In: Proceedings of the XXVIII Congreso Anual de la Sociedad Española de Ingeniería Biomédica, Madrid, Spain.

[19] J. Bernal, J. Sánchez, and F. Vilarino, A region segmentation method for colonoscopy images using a model of polyp appearance, in Pa ern Recognition and Image Analysis, Springer, pp. 134-142, 2011.

[20] J. Bernal, J. Sánchez, and F. Vilarino, Towards automatic polyp detection with a polyp appearance model, Pattern Recognit., vol. 45, no. 9, pp. 3166-3182, 2012.

[21] R. Dahyot, F. Vilariño, and G. Lacey, Improving the quality of color colonoscopy videos, J. Image Video Process, vol. 2008, p. 15, 2008.

[22] M. Arnold, S. Ameling, A. Ghosh, and G. Lacey, Quality Improvement of Endoscopy Videos, in Proceedings of the 8th IASTED International Conference on Biomedical Engineering, Innsbruck, Austria, 2011.

Chapter 6

Robotic Splenic Flexure and Transverse Colon Resections

Igor Monsellato, Maria Caterina Canepa,
Vittorio d'Adamo, Giuseppe Spinoglio,
Fabio Priora and Luca Matteo Lenti

Additional information is available at the end of the chapter

http://dx.doi.org/10.5772/intechopen.69881

Abstract

Since the 1990s, laparoscopic technique has become a standard approach for several surgical procedures in the field of colorectal surgery. Laparoscopic approach to splenic flexure and transverse colon cancer, however, is still a matter of debate and considered challenging for both anatomical and technical aspects. The relationship with the spleen and the absence of a consensus on the extent of surgery for splenic flexure cancer are two of several aspects that make splenic flexure surgery mostly debated. Robotic technique has overcome some pitfalls of laparoscopy, thanks to its stability of vision, tremor filtering, and fine movements of the robotic arms that can help in better identifying and managing both vascular structures and side organs, thus avoiding splenic and pancreatic injuries. In addition, robotic system can allow a better fashioning of the intracorporeal anastomosis, and the advent of fluorescence is useful to guide dissection and to evaluate the vascularization of the colon. Herein we discuss a standardized approach for robotic splenic flexure resection and transverse colon.

Keywords: splenic flexure cancer, transverse colon cancer, robotic colorectal surgery, robotic technique, robotic splenic flexure resection, robotic transverse colon resection, robotic intracorporeal anastomosis

1. Introduction

Since the 1990s, laparoscopic technique has become a standard approach for several surgical procedures in the field of colorectal surgery [1]. All the main prospective trials comparing open and laparoscopic technique for colorectal cancer have shown same clinical and oncological outcomes of the two approaches [2–5]. Laparoscopic approach to splenic flexure and transverse colon cancer, however, has not been investigated and it is still a matter of debate, mainly due to the rare incidence of the cancer of the left flexure, ranging approximately from 3 to 10% of all colon cancers, and to technical difficulties in approaching the transverse colon. Splenic flexure cancers are generally considered as all those cancers occurring between the distal part of the transverse colon and the proximal part of the descending colon [6]. Pure transverse colon cancers are commonly defined as all those cancers occurring in the middle part of the transverse colon. Surgical technique for this kind of tumors is not standardized yet, because of anatomical aspects and technical issues. Laparoscopic approach has been considered a challenging procedure, with longer operative time than in open surgery and a relative risk of splenic and pancreatic injuries, suggesting its use by expert surgeons and for early stage disease. Robotic surgery has been introduced in colorectal surgery about 15 years ago, and it is spreading worldwide, thanks to its advantages over laparoscopic technique.

2. Robotic splenic flexure resection

2.1. Patient positioning, robot docking, and operating theatre setup

2.1.1. Operative setting overview

Patient is placed in the reverse Trendelenburg position (15°), with a 30°-tilt to the right with the arms alongside the trunk and the legs abducted (**Figure 1**). The cart approaches the operative table from patient's left hip (**Figure 2**). The procedure is carried out with a five-trocar technique and begins with the insertion of the Veress needle in the left hypochondrium and the induction

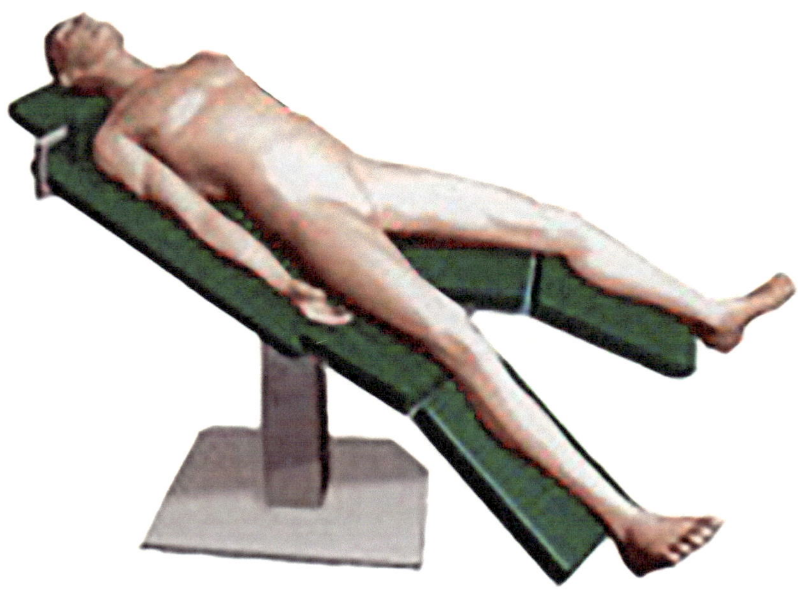

Figure 1. Position of the patient on the operative table.

http://dx.doi.org/10.5772/intechopen.69881

of a 12-mmHg pneumoperitoneum. The optical trocart is placed 2 cm right and up the umbilical scar. The robotic trocarts R1, R2, and R3 are inserted in the right iliac fossa at the cross between the line passing through the antero-superior iliac spine and the umbilical scar, and the middle clavicular line; in the epigastrium/left flank between the midline and the left-middle clavicular line, and in the right hypochondrium 2 cm below the right rib margin along the middle clavicular line, respectively. A laparoscopic 12-mm trocart is placed in the right flank between R1 and R3, for the assistant at the operative table. Arm 1 is connected to R1, arm 2 is connected to R2, and arm 3 is connected to R3. The complete trocarts and operating theatre setups are shown in **Figures 3** and **4**.

2.1.2. Robotic instruments and setting

Robotic instruments used in this procedure are bipolar fenestrated forceps, for coagulation and traction, the ProGrasp for traction and exposure, and robotic scissors for cutting and blunt dissection (when used with closed jaws). The robotic monopolar scissors are mounted on arm 1, the robotic bipolar fenestrated forceps on arm 3, and the ProGrasp on arm 2 (**Figure 5**).

2.2. Exploration of the abdominal cavity, intraoperative hepatic ultrasonography and mobilization of the splenic flexure

A laparoscopic exploration of the abdominal cavity and an intraoperative ultrasonography of the liver are systematically performed to identify the site of the neoplasm (tattoo or the cancer itself) and to complete the staging of the disease. This is a fundamental step that allows also in finding out the connections between the splenic flexure and the inferior pole of the spleen. The

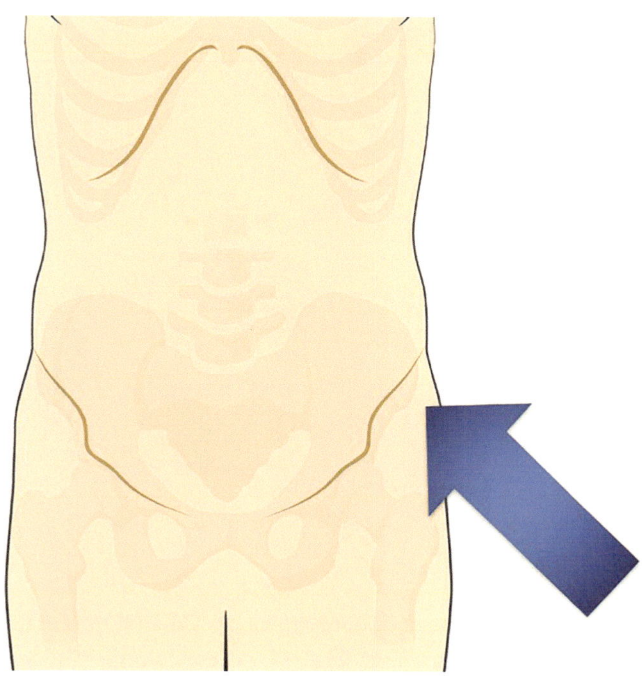

Figure 2. Direction of the docking of the robotic cart in left colic flexure resection.

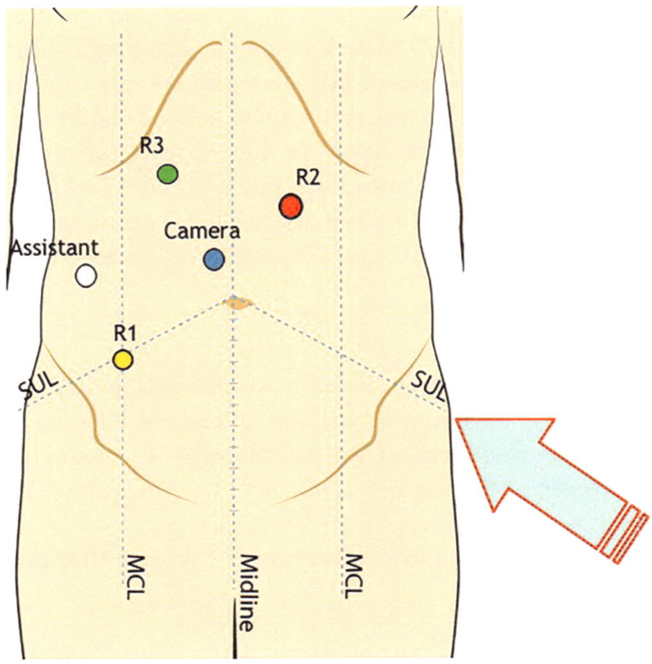

Figure 3. Trocarts position in left colic flexure resection. SUL, spine-umbilical line; MCL, middle clavicular line.

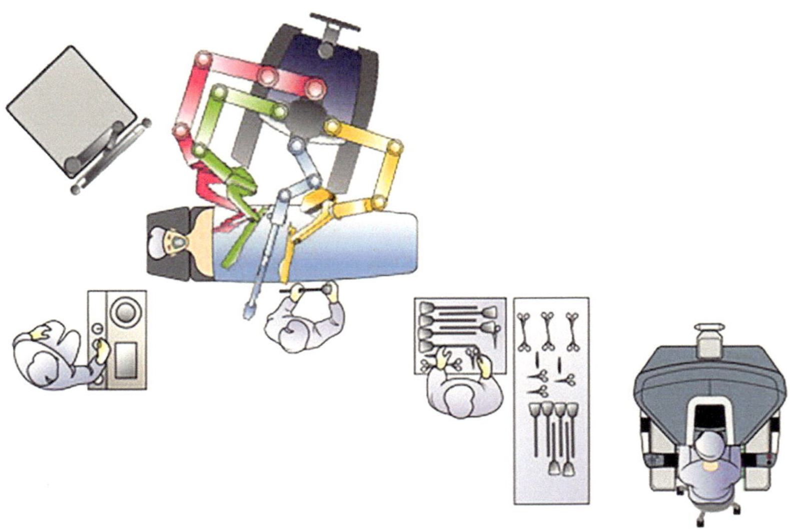

Figure 4. Operative theater setting.

robotic arms are connected to the trocarts (robot docking). The first step of the flexure takedown is the dissection of the gastrocolic ligament. The transverse colon is pulled down by the assistant with a laparoscopic grasper, while the stomach is pulled up by the bipolar forceps on arm 3, in order to maximize the exposition of the gastrocolic ligament and to identify the Bouchet's

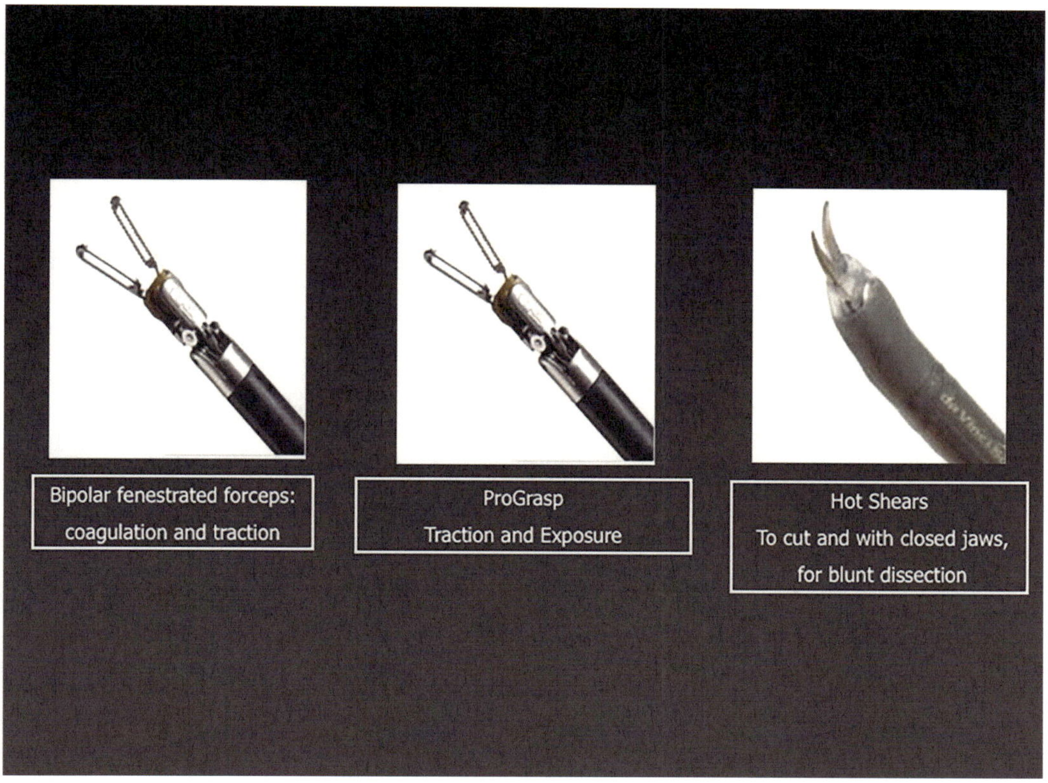

Figure 5. Robotic instruments used during robotic left colic flexure resection.

area, the starting point of the dissection carried out by the robotic scissors on arm 1 in a right-to-left direction (**Figure 6**). The dissection continues till the lower pole of the spleen is reached, then the splenocolic ligament and the superior part of the left paracolic gutter are incised. The access to the lesser sac is achieved. The inferior margin of the pancreas is identified and the root of the mesocolon is incised 1 cm below the pancreatic margin by the robotic scissors on arm 1, from left to right, till reaching the first jejunal loop, at the Treitz area. The transverse and left colon are medialized by the assistant and the separation of the Toldt's fascia from the Gerota's fascia is carried out in a lateral-to-medial direction; during this step, the paracolic gutter is completely incised up to the sigmoid colon. The takedown of the splenic flexure is completed.

2.3. Vascular dissection and lymphadenectomy

2.3.1. Brief summary of the vascular and lymphatic drainage anatomy of the left colic flexure

Vascular anatomy of the left colic flexure is constituted by secondary branches of the two main intestinal vascular trunks; blood supply is provided by the left branch of the middle colic artery originating from the superior mesenteric artery, and by the left colic artery (LCA), originating from the inferior mesenteric artery (IMA); venous drainage flows into the superior mesenteric vein, through the left branch of the middle colic vein, and into the inferior mesenteric vein (IMV), through the left colic vein (LCV).

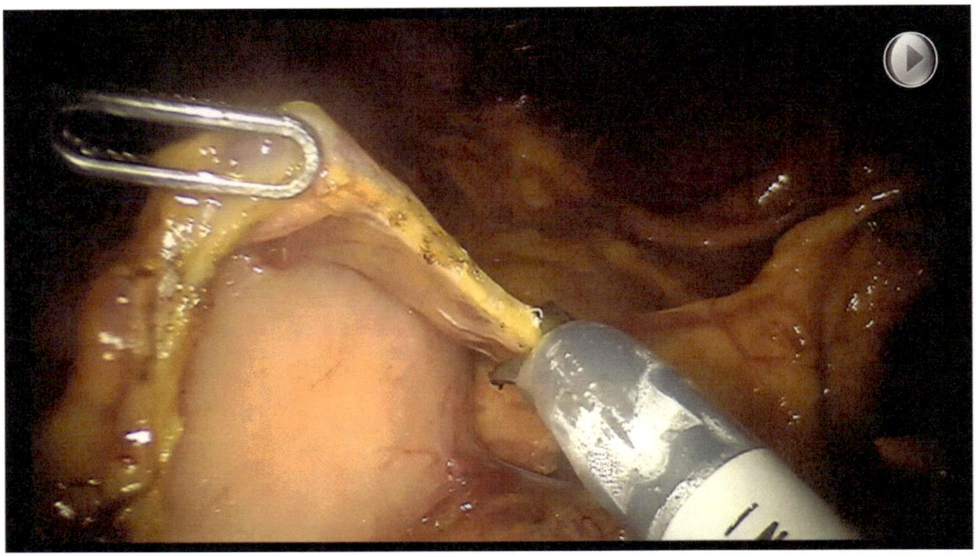

Figure 6. Dissection of the gastrocolic ligament. The transverse colon is pulled down by the assistant with a laparoscopic grasper, while the stomach is pulled up by the bipolar forceps on arm 3, in order to maximize the exposition of the gastrocolic ligament and to identify the Bouchet's area.

Splenic flexure cancer has various lymphatic drainage pathways. The standard lymphatic way is satellite to the left branch of the middle colic artery and left colic artery, but lymphatic metastases to the infrapancreatic lymph node region and the splenic hilum have been reported. Indocyanine green (ICG) fluorescence may help analyzing metastatic lymphatic spread, if injected subserosally or submucosally. The optimal dose range is between 0.1 and 0.5 mg/kg and should not exceed 2 mg/kg. For the detection of the lymph flow, a dose of ICG of 2.5 mg/1.0 mL is injected into the subserosal-submucosal layer around the tumor at two points after trocar insertion; the lymph flow is observed using the robotic integrated near-infrared system (NIR) 30 min after ICG injection.

2.3.2. Description of the vascular dissection

The transverse colon is pulled upward by the assistant with the laparoscopic grasper, and the left colon is lifted up and laterally by the ProGrasp on arm 2. The inferior mesenteric vein (IMV) is identified at the inferior margin of the pancreas. The dissection starts at the lateral margin of the IMV in order to identify the left colic vein (LCV). An accurate lymphadenectomy of the root of the IMV is performed. The left mesocolon is lifted up by the ProGrasp on arm 2, while the IMV is medialized by the assistant with a laparoscopic grasper. The dissection continues till the LCV is identified and isolated between non-adsorbable clips, applied by the assistant or by the robotic clip applier on arm 1, and cut by the assistant or by the scissors on arm 1 (**Figure 7a** and **b**). The dissection is carried out with a medial to lateral direction, joining the previous plane between Toldt's and Gerota's fascia. The sigmoid colon is completely mobilized, preserving the left gonadal vessels and the ureter, lying down Gerota's fascia (**Figure 8**).

The left mesocolon is then lifted up by the ProGrasp on arm 2 to identify the inferior mesenteric artery (IMA): the dissection follows the lateral aspect of the IMA till reaching the origin

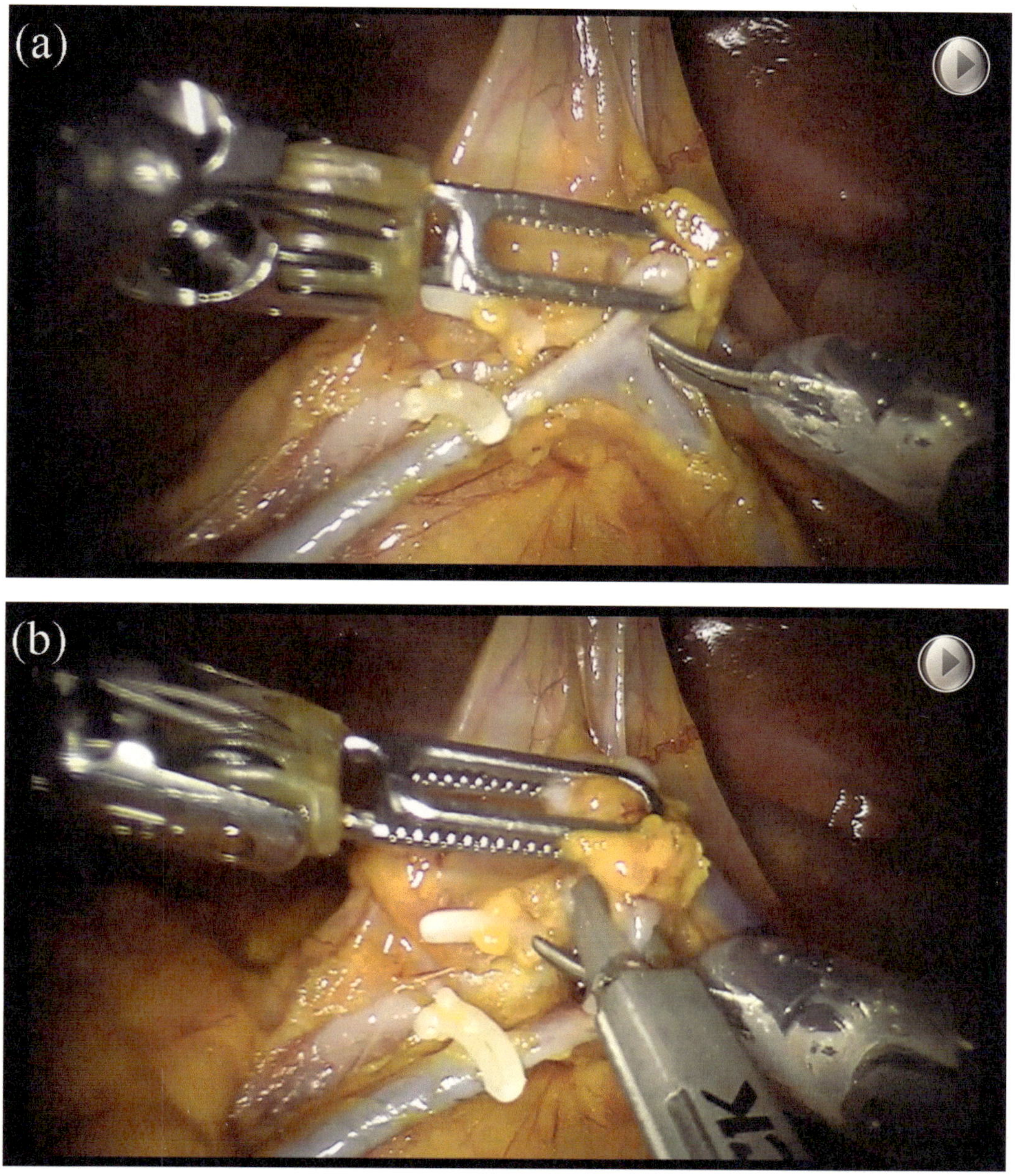

Figure 7. Vascular dissection. the LCV is identified and isolated between non-adsorbable clips, applied by the assistant or by the robotic clip applier on arm 1, and cut by the robotic scissors on arm 1. LCV, left colic vein.

of the left colic artery (LCA). An accurate lymphadenectomy of the origin of the IMA is performed. The LCA is isolated between non-adsorbable clips and cut by the assistant or by the robotic scissors on arm 1 (**Figure 9**).

The vascular dissection continues with the isolation of the left branches of the middle colic vessels (LMCV) (**Figure 10a**). The transverse mesocolon is pulled upward by the ProGrasp on arm 2,

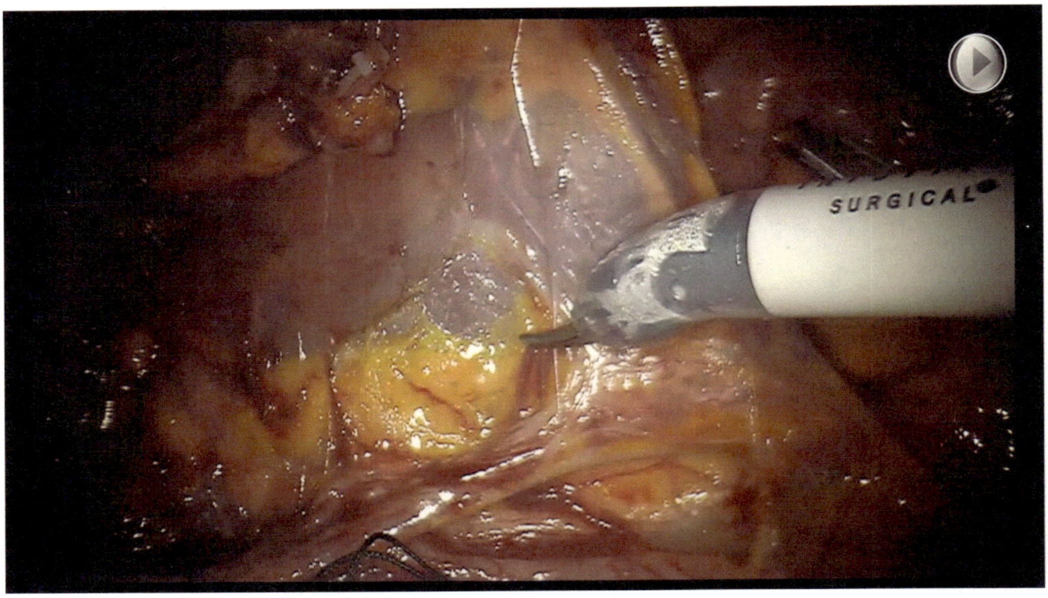

Figure 8. Mobilization of the sigmoid colon. The left gonadal vessels and ureter are preserved under Gerota's fascia.

and the identification of the main trunk of the middle colic vessels starts at its origin from the superior mesenteric vein (SMV), upward. The dissection is carried out by the robotic bipolar forceps on arm 3 and the robotic scissors on arm 1. After identification of the main trunk, the left branch is dissected and freed from the surrounding lymphatic and fatty tissue, and cut by the assistant or by the scissors on arm 1, after being isolated between clips, as well (**Figure 10b** and **c**).

Lymphadenectomy and vascular dissection have been completed.

2.4. Transverse and left colon transection, anastomosis

Once the mobilization of the splenic flexure and the vascular dissection are completed, the evaluation of the vascularization of the colon with ICG is performed to identify the correct site of transection (**Figure 11a**). After intravenous injection, in a time interval between 5 and 30 s, ICG reaches the arterial and venous vessels. The assistant, then, cut the transverse colon and the proximal sigmoid colon by a laparoscopic linear stapler (**Figure 11b**). A robotic linear stapler can be used if available on arm 1. A recheck of both the two colonic stumps is carried out to avoid postoperative risk of anastomotic or stumps dehiscence, mainly caused by tissue devascularization (**Figure 11c** and **d**). The specimen is inserted into an endobag for further removal. The two colonic stumps are approached. The robotic bipolar forceps on arm 3 holds the descending colon stump and a colotomy is performed at the level of the tenia, with the robotic scissors on arm 1, as well as for the transverse colon stump. The laparoscopic linear stapler is introduced into both the two colostomies while the surgeon at the console helps the introduction of the two branches of the stapler inside the colonic stumps with the robotic bipolar forceps on arm 3. A colocolic side-to-side antiperistaltic mechanical anastomosis is then

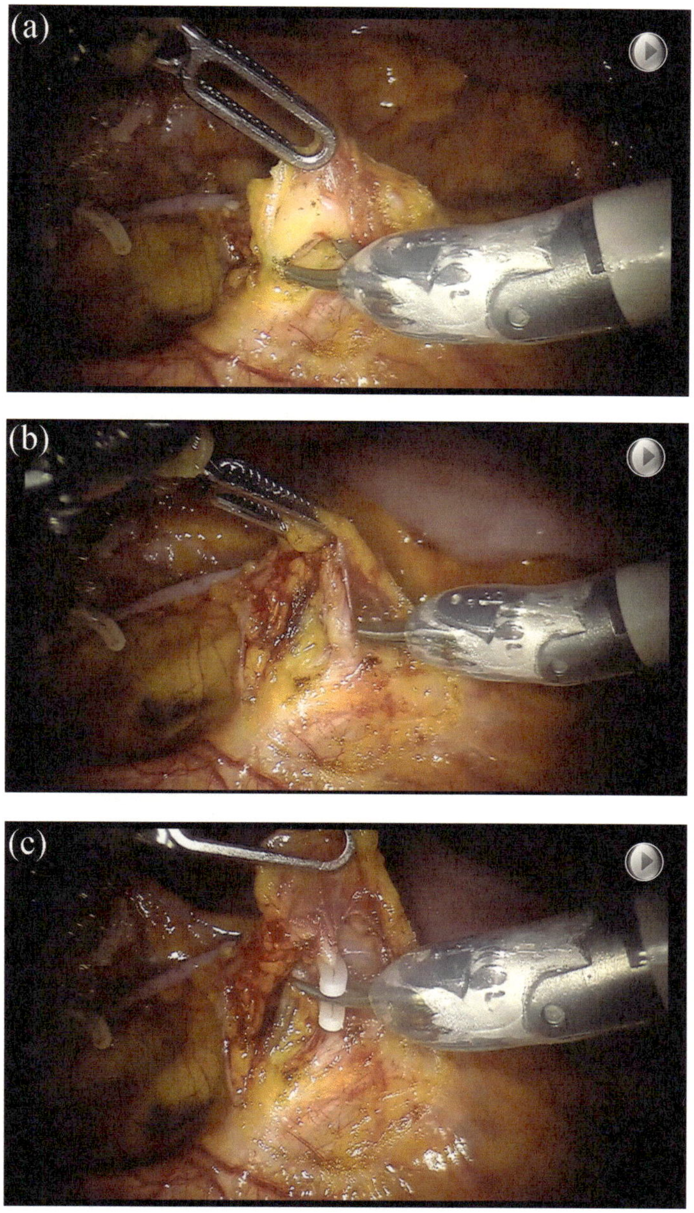

Figure 9. Vascular dissection. The LCA is isolated between non-adsorbable clips and cut by the robotic scissors on arm 1.

performed (**Figure 12**). The entry hole of the stapler is closed by two running barbed sutures starting from the opposite angles. The first running suture is performed from the inferior angle upward. The second suture is performed from the upper angle downward (**Figure 13a** and **b**). Afterwards, the robotic system is undocked and a Pfannenstiel incision is performed for specimen extraction.

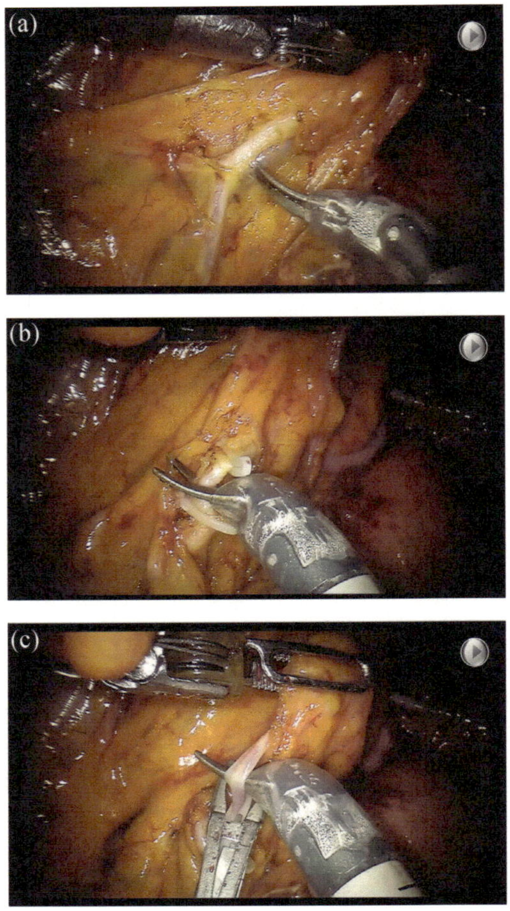

Figure 10. Vascular dissection. Identification of the left middle colic vessels (a), isolation between clips (b) and section (c).

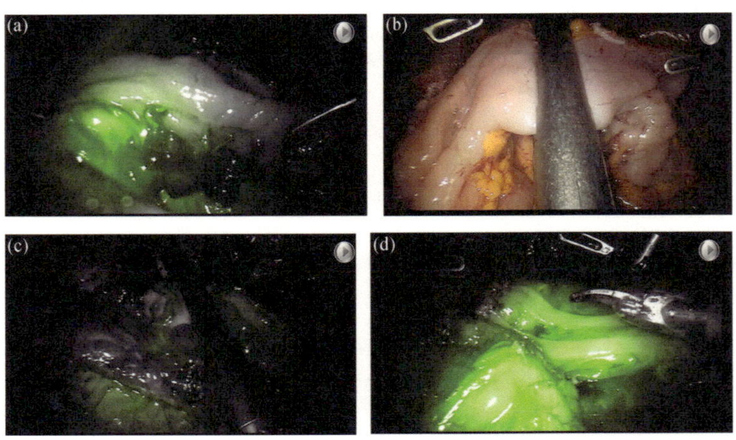

Figure 11. Evaluation of the vascularization of the colon with ICG (a), section of the colon (b), recheck of the two colonic stumps (c-d).

http://dx.doi.org/10.5772/intechopen.69881

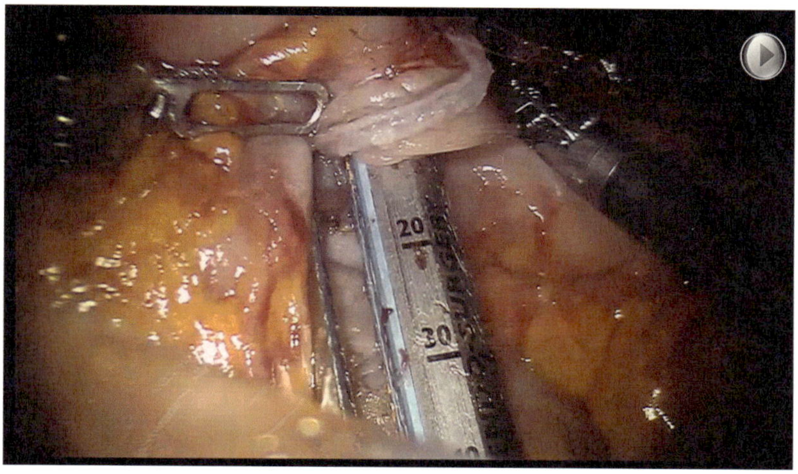

Figure 12. Fashioning of the colo-colic side-to-side mechanical anastomosis.

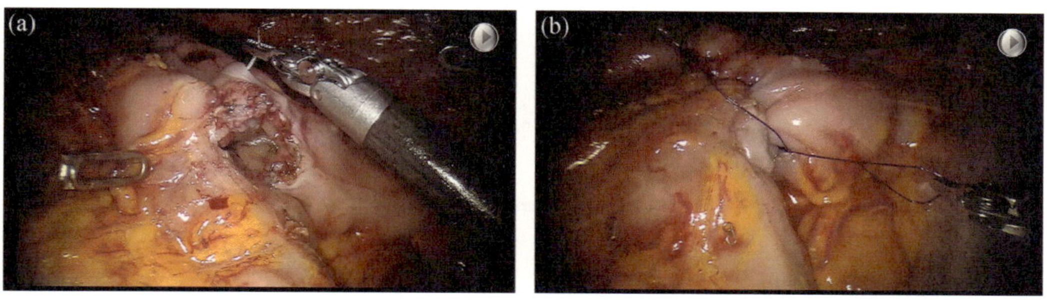

Figure 13. Closure of the entry hole.

3. Transverse colon resection

3.1. Patient positioning, robot docking, and operating theatre setup

3.1.1. Operative setting overview

Patient is placed in anti-Trendelenburg position with the arms along the trunk and the legs abducted. The robotic cart approaches the operative table from patient's head (**Figure 14**). The procedure is carried out with a five-trocar technique and begins with the insertion of the Veress needle in the left hypochondrium and the induction of a 12-mmHg pneumoperitoneum. The optical trocart is placed 2 cm right the umbilical scar. Three robotic trocarts are placed in the right (R2) and left (R1) hypochondrium 2–3 cm under the rib margin and in the right flank (R3) along the middle clavicular line 2 cm below the transverse umbilical line. A laparoscopic 12 mm trocart is inserted in the left flank along the middle clavicular line 2 cm

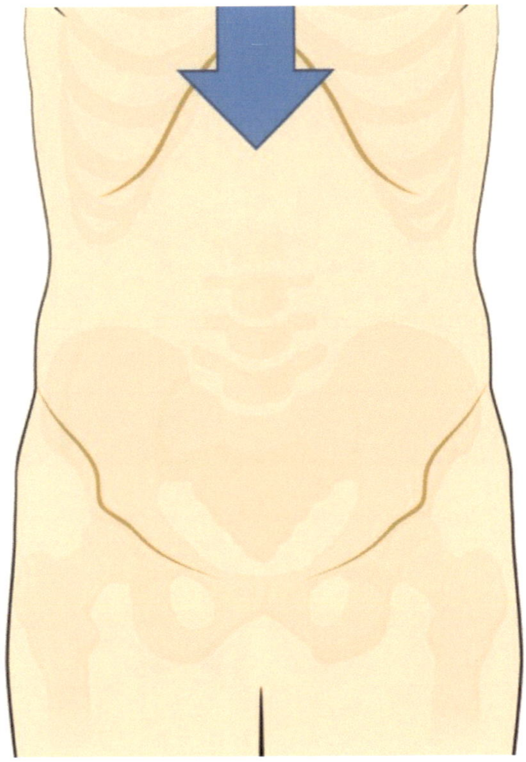

Figure 14. Direction of the docking of the robotic cart in transverse colon resection.

below the transverse umbilical line for the assistant. Arm 1 is connected to R1, arm 2 is connected to R3, and arm 3 is connected to R2. The complete trocarts and operating theatre setup are shown in **Figures 15** and **16**.

3.1.2. Robotic instruments and setting

As for splenic flexure procedure, robotic instruments used in transverse colon resection are bipolar fenestrated forceps, for coagulation and traction, the ProGrasp for traction and exposure, and robotic scissors for cutting and blunt dissection (when used with closed jaws). The robotic monopolar scissors are mounted on arm 1, the robotic bipolar fenestrated forceps on arm 2, and the ProGrasp on arm 3.

3.2. Vascular dissection and lymphadenectomy

3.2.1. Brief summary of the vascular and lymphatic drainage anatomy of the transverse colon

Transverse colon receives blood supply from the two main intestinal trunks: the superior mesenteric artery (SMA) and the IMA, via the middle colic arteries and the left colic artery, as well as the venous drainage is tributary of both the two vascular systems (SMV and IMV). Venous drainage, however, is especially variable and closely related to pancreatic and omental

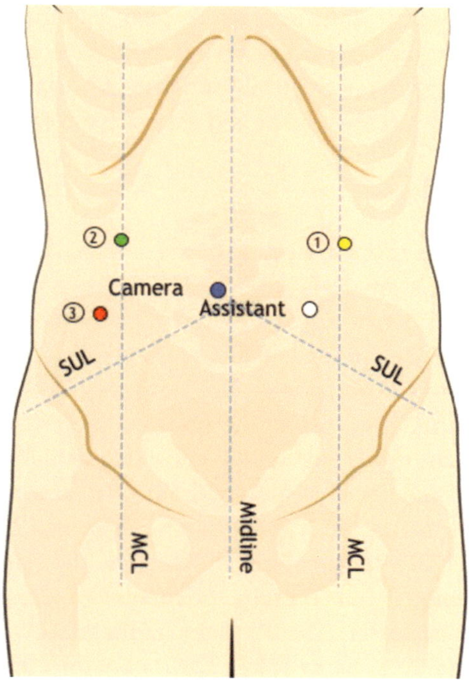

Figure 15. Trocarts position in transverse colon resection. SUL, spine-umbilical line; MCL, middle clavicular line.

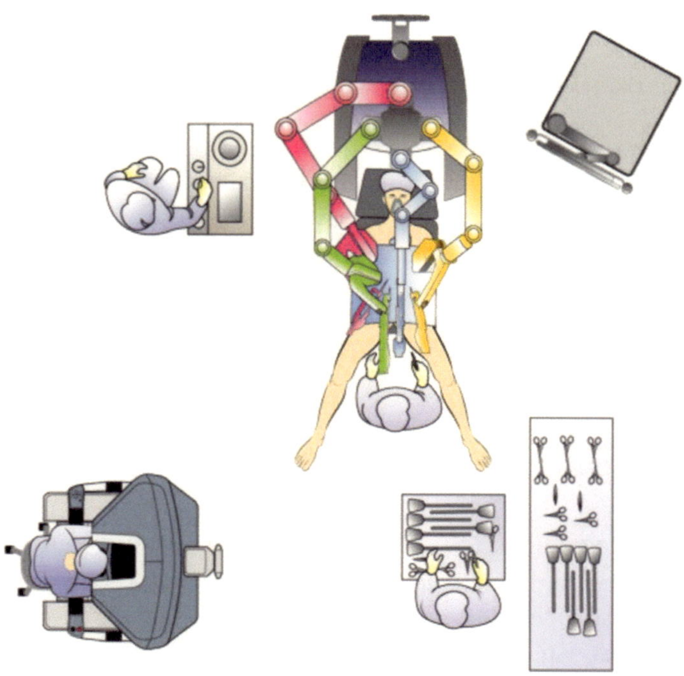

Figure 16. Operative theater setting.

veins, as for the close relationship of the transverse colon with the greater omentum, the pancreas, and the stomach. Even though locoregional lymphadenectomy of the root of the SMV and SMA is commonly considered oncologically adequate, some authors suggest exploring and dissecting lymph nodes of the infrapancreatic and gastroepiploic region.

3.2.2. Description of the vascular dissection

The first step of the procedure is the dissection of the gastrocolic ligament carried out by the robotic scissors on arm 1. The stomach is pulled up by the surgeon with the robotic ProGrasp on arm 3, and the transverse colon is pulled downward by the assistant. The dissection continues laterally to the sigmoid colon on the left side to the cecum on the right: the phrenicocolic and splenocolic ligament, and the parietocolic ligament are sectioned on the left; on the right, the gastrocolic ligament is dissected below the gastroepiploic vessels, performing a locoregional lymphadenectomy. The right colon is then retracted medially by the assistant and by the robotic ProGrasp on arm 3, and the dissection of the right parietocolic ligament is performed by the robotic scissors on arm 1. A blunt dissection is performed lateral-to-medial from both sides and over the pancreas till reaching the Treitz area and the origin of the middle colic vessels. Then, the transverse colon is pulled up by the robotic ProGrasp on arm 3, enhancing the main trunk of the middle colic vessels. The dissection of the root of the transverse mesocolon is completed toward the end of the pancreatic tail. The root of the main trunk of the middle colic vessels is clipped by the assistant or by the robotic clip applier and sectioned by the assistant or by the robotic scissors on arm 1, and the locoregional lymphadenectomy is carried out. The incision of the transverse mesocolon is performed by the robotic scissors, starting from the middle colic vessels straight to the flexure on both sides.

3.3. Transection of the transverse colon and anastomosis

The transection of the colon is performed by the assistant with a laparoscopic flexible stapler or by the surgeon with the robotic stapler on arm 1, and it includes both the two flexures. The left and the right colon are joined and a colotomy is carried out at the closed margin of both the two colonic stumps. The right colon is held by the surgeon with the ProGrasp on arm 2, the left colon by the assistant. A double running suture colocolic end-to-end anastomosis is performed with two needle-holders on arms 1 and 3. Afterward, the robotic system is undocked and a Pfannenstiel incision is performed for specimen extraction.

4. Generic considerations, technical review, and future perspectives

The role of minimally invasive surgery has been recently established in the colorectal field thanks to a series of randomized clinical trials that compared laparoscopic and open techniques. Their results definitely eliminated any doubts concerning the oncological adequacy of minimally invasive treatment. The spread of minimally invasive surgery may be also justified by the reduced postoperative pain, decreased hospital stay and faster postoperative recovery, reduced incidence of postoperative complications, improved cosmetic outcome, and decreased incidence of incisional hernias. Alongside these encouraging results, technologic innovations have been introduced in minimally invasive surgery. Robotic technique has spread worldwide thanks to its advantages over standard and advanced laparoscopy (three-dimensional (3D)). Several generic aspects of robotic assistance, such as three-dimensional view, better ergonomics, magnified vision, and articulated tips of the robotics instruments are reported to be significant technical advantages in colorectal surgery. Splenic flexure and transverse colon resections seem to be challenging procedures, which robotics may help with. Surgical approach of left flexure and transverse colon cancers is not standardized yet, due to the rare incidence of the flexure cancer and the technical difficulties of the latter. Left colic flexure cancers have an incidence ranging from about 3–10% of all colon cancers [7] and were initially correlated to a poor prognosis and a high risk of obstruction [8]. This correlation has been recently overcome by some studies showing comparable survival outcomes to those of other colonic cancers, and demonstrating that neither the splenic flexure site nor colonic obstruction has an independent influence on patient survival after surgery [8–10]. One of the aspects that remain controversial is the extension of lymphadenectomy [6, 11]. As previously reported, the classic pathways follow the main nourishing arterial trunks: the middle colic vessels, essentially the left branch, and the left colic artery. Some authors argue that lymphatic spread may follow the IMV and the IMA, thus requiring their ligation and a consequent standard left colectomy in order to achieve an oncologically adequate lymphadenectomy; aberrant metastatic pathways to the infrapancreatic lymph node region and the splenic hilum have been reported [12], even though no systematic data in the literature regarding the frequency of lymphatic drainage roots at this site have been clarified yet. Recently, indocyanine green fluorescent imaging (ICG) in colorectal cancer has been used to evaluate the blood flow, but there are few reports on the lymphatic flow [13–15]. Some authors conducted a study on the pattern of lymph flow for splenic flexure colon cancers with ICG on 31 consecutive patients [15]. The amount of ICG injected was 2.5 mg (1 mL of solution) into the subserosal-submucosal layer. The main lymphatic diffusion was observed through the IMV and LCA areas, with or without the presence of aberrant vascularization. The conclusion was that lymph node dissection of the root of the IMV area is important and it should be always performed, avoiding ligation of both the left middle colic artery (lt-MCA) and LCA, in those cases without widespread lymph node metastases. Unnecessary splenectomy is one of the main complications reported on laparoscopic studies for splenic flexure colon cancers, due to the anatomical relations between these two organs, and the characteristics of the laparoscopic instruments. Poor dexterity, instrument stiffness, and a limited range of motion make splenic flexure resection a challenging procedure, requiring also several modification of patient's position. Moreover, the use of 3D laparoscopy is controversial as it seems to fail showing any advantages in colonic resection or other more complex procedures, as it is in its infancy and further comparative studies are necessary to assert whether it can reduce learning curve [16–18]. Robotic assistance may help performing an accurate lymphadenectomy thanks to motion scaling, tremor filtering, 7-degrees of freedom and the 3D magnified view, avoiding unnecessary vessel ligation or inadverted injuries to the surrounding organs. These results suggest that robotic assistance, associated to ICG imaging, may introduce the concept of "tailored" surgery and can facilitate surgical resection of splenic flexure colon cancer. Further studies on lymph flow pattern may lead to a "standardization" of this procedure. Fluorescence was integrated into the da Vinci Si HD System (Intuitive Surgical, Sunnyvale, CA,

USA) in 2010. The surgeon can quickly switch between normal viewing mode to fluorescence (near-infrared light) by pressing the pedal on the surgical console (**Figure 17**). Indocyanine green is a sterile, water-soluble protein-binding dye with low toxicity and fast biliary excretion. ICG fluorescence imaging system is a simple, safe, useful method and can be used in several fields of general surgery, particularly in oncologic surgery [13].

Laparoscopic transverse colon resection for middle cancers is a rare procedure as population suffering with it is too small. Moreover, transverse colectomy requires advanced laparoscopic surgical skills and, consequently, a longer learning curve than other colorectal procedures. Dissection of the middle colic vessels and locoregional lymphadenectomy are more challenging than in other laparoscopic colectomies as well as the complete mobilization of hepatic and splenic flexures, which is an essential step of transverse colectomy. Colic flexures takedown may help anastomosis fashioning; even though there is no statistical difference between the advantages of intracorporeal versus extracorporeal anastomosis, it is preferable to perform an intracorporeal colocolic end-to-end anastomosis because of some well-known advantages: better chance to choose the site of the minilaparotomy (suprapubic or median), especially in obese patients, low traction on the mesentery and avoidance of twisting of the mesentery [19–21]. Some authors argue that excessive mobilization of the colon, without the flexures

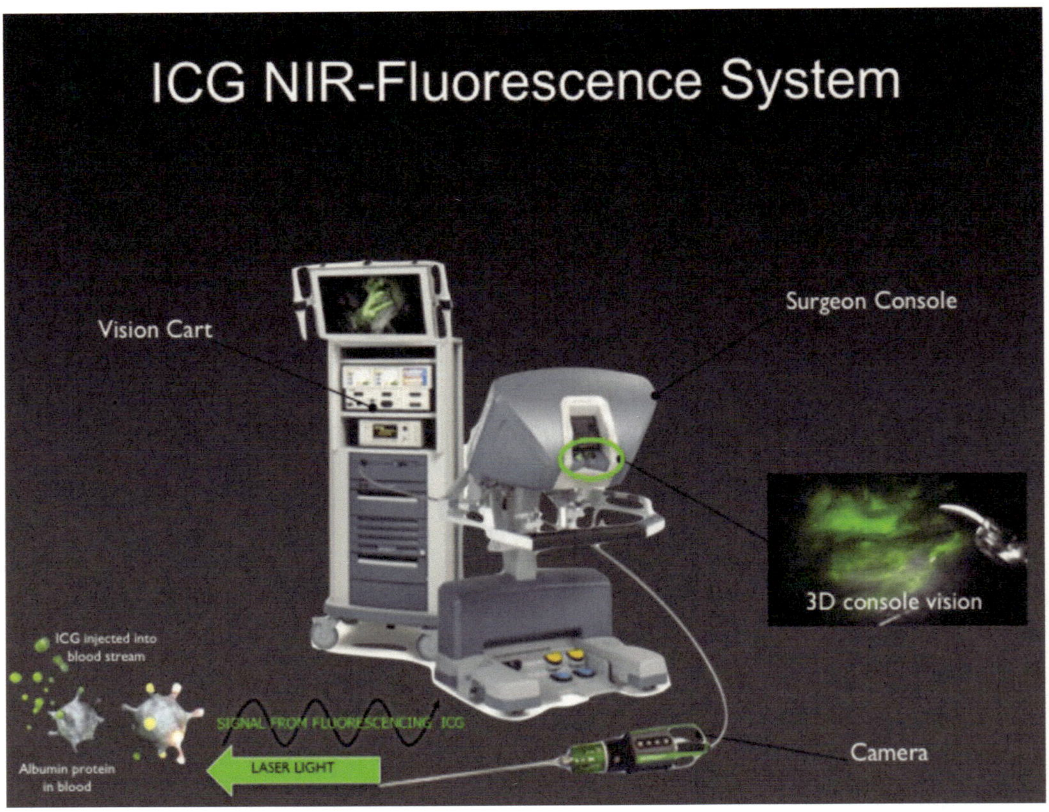

Figure 17. ICG NIR-Fluorescence System. Fluorescence was integrated into the da Vinci Si HD System (Intuitive Surgical, Sunnyvale, CA, USA) in 2010. The surgeon can quickly switch between normal viewing mode to fluorescence (near-infrared light) by pressing the pedal on the surgical console.

http://dx.doi.org/10.5772/intechopen.69881

takedown, can be avoided when adopting an intracorporeal anastomosis [22], but it can result in a unadverted traction on the anastomosis itself and a moderate risk of dehiscence in the postoperative, then, it is always preferable to take down both the two flexures even in an intracorporeal anastomosis. Robotic approach provides specific advantages in intracorporeal anastomosis sewing, thanks to the endo-wrist function and the stability of the robotic arms, thus reproducing all the steps as in open surgery, and the 3D magnified view. Initially, the lack of tactile sensation was considered a pitfall of the robotic system, and several studies of engineering are still ongoing in order to provide a tactile sensation by the robotic system. This aspect, however, was recently confuted as it was shown that visual feedback of an expert surgeon can successfully replace tactile sensation, without the need of a tactile-feedback device [23].

In conclusion, few cases have been reported on robotic splenic flexure and transverse colon resection, but robotic assistance seems to provide several advantages on performing these procedures. Further studies are necessary to assess the real role of robotics in the treatment of the splenic flexure and mid-transverse colon cancers.

Author details

Igor Monsellato[1]*, Maria Caterina Canepa[1], Vittorio d'Adamo[1], Giuseppe Spinoglio[2], Fabio Priora[1] and Luca Matteo Lenti[1]

*Address all correspondence to: imonsellato@libero.it

1 Azienda Ospedaliera Nazionale SS. Antonio e Biagio e Cesare Arrigo, Alessandria, Italy

2 EIO – European Institute of Oncology, Milan, Italy

References

[1] Jacobs M, Verdeja JC, Goldstein HS. Minimally invasive colon resection (laparoscopic colectomy). Surgical Laparoscopy & Endoscopy. 1991;**1**:144-150

[2] Lacy AM, Garcia-Valdecasas JC, Delgado S, Castells A, Taura P, Pique JM, Visa J. Laparoscopic-assisted colectomy versus open colectomy for treatment of non-metastatic colon cancer: A randomised trial. Lancet. 2002;**359**(9325):2224-2229

[3] Fleshman J, Sargent DJ, Green E, Anvari M, Stryker SJ, Beart Jr RW, Hellinger M, Flanagan Jr R, Peters W, Nelson H; Clinical Outcomes of Surgical Therapy Study Group. Laparoscopic colectomy for cancer is not inferior to open surgery based on 5-year data from the COST Study Group trial.. Annals of Surgery. 2007;**264**(4):655-662

[4] Green BL, Marshall HC, Collinson F, Quirke P, Guillou P, Jayne DG, Brown JM. Long-term follow-up of the Medical Research Council CLASICC trial of conventional versus laparoscopically assisted resection in colorectal cancer. British Journal of Surgery. 2013;**100**(1):75-82

[5] Colon Cancer Laparoscopic or Open Resection Study Group, Buunen M, Veldkamp R, Hop WC, Kuhry E, Jeekel J, Haglind E, Påhlman L, Cuesta MA, Msika S, Morino M, Lacy A, Bonjer HJ. Survival after laparoscopic surgery versus open surgery for colon cancer: Long-term outcome of a randomised clinical trial. Lancet Oncology. 2009;**10**(1):44-52

[6] Steffen C, Bokey EL, Chapuis PH. Carcinoma of the splenic flexure. Diseases of the Colon & Rectum. 1987;**30**(11):872-874

[7] Kim CW, Shin US, Yu CS, Kim JC. Clinicopathologic characteristics, surgical treatment and outcomes for splenic flexure colon cancer. Cancer Research and Treatment. 2010;**42**:69-76

[8] Levien DH, Gibbons S, Begos D, Byrne DW. Survival after resection of carcinoma of the splenic flexure. Diseases of the Colon & Rectum. 1991;**34**:401-403

[9] Secco GB, Ravera G, Gasparo A, Percoco P, Zoli S. Segmental resection, lymph nodes dissection, and survival in patients with left colon cancer. Hepatogastroenterology. 2007;**54**(74):422-426

[10] Nakagoe T, Sawa T, Tsuji T, Jibiki M, Nanashima A, Yamaguchi H, Yasutake T, Ayabe H, Ishikawa H. Carcinoma of the splenic flexure: multivariate analysis of predictive factors for clinicopathological characteristics and outcome after surgery. Journal of Gastroenterology. 2000;**35**:528-535

[11] Nakagoe T, Sawai T, Tsuji T, Jibiki M, Ohbatake M, Nanashima A, Yamaguchi H, Yasutake T, Kurosaki N, Ayabe H, Ishikawa H. Surgical treatment and subsequent outcome of patients with carcinoma of the splenic flexure. Surgery Today 2001;**31**(3):204-209

[12] Perrakis A, Weber K, Merkel S, Matzel K, Agaimy A, Gebbert C, Hohenberger W. Lymph node metastasis of carcinomas of transverse colon including flexures. Consideration of the extramesocolic lymph node stations. International Journal of Colorectal Disease. 2014;**29**(10):1223-1229

[13] Marano A, Priora F, Lenti LM, Ravazzoni F, Quarati R, Spinoglio G. Application of fluorescence in robotic general surgery: Review of the literature and state of the art. World Journal of Surgery. 2013;**37**(12):2800-2811. DOI: 10.1007/s00268-013-2066-x

[14] Nishigori N, Koyama F, Nakagawa T, Nakamura S, Ueda T, Inoue T, Kawasaki K, Obara S, Nakamoto T, Fujii H, Nakajima Y. Visualization of lymph/blood flow in laparoscopic colorectal cancer surgery by ICG fluorescence imaging (lap-IGFI). Annals Surgery of Oncology. 2016;**23**(Suppl 2):266-274

[15] Watanabe J, Ota M, Suwa Y, Ishibe A, Masui H, Nagahori K. Evaluation of lymph flow patterns in splenic flexural colon cancers using laparoscopic real-time indocyanine green fluorescence imaging. International Journal of Colorectal Disease. 2017;**32**(2):201-207

[16] Currò G, Cogliandolo A, Bartolotta M, Navarra G. Three-Dimensional versus Two-Dimensional laparoscopic right hemicolectomy. Journal of Laparoendoscopic & Advanced Surgical Techniques. Part A. 2016;**26**(3):213-217. DOI: 10.1089/lap.2015.0557. Epub 2016 Feb 10

[17] Currò G, La Malfa G, Caizzone A, Rampulla V, Navarra G. Three-dimensional (3D) versus two-dimensional (2D) laparoscopic bariatric surgery: A single-surgeon prospective randomized comparative study. Obesity Surgery. 2015;**25**(11):2120-2124

[18] Currò G, La Malfa G, Lazzara S, Caizzone A, Fortugno A, Navarra G. Three-dimensional versus two-dimensional laparoscopic cholecystectomy: Is surgeon experience relevant? Journal of Laparoendoscopic & Advanced Surgical Techniques. Part A. 2015;**25**(7):566-570. DOI: 10.1089/lap.2014.0641. Epub 2015 Jun 15

[19] D'Annibale A, Pernazza G, Morpurgo E, Monsellato I, Pende V, Lucandri G, Termini B, Orsini C, Sovernigo G. Robotic right colon resection: Evaluation of first 50 consecutive cases for malignant disease. Annals Surgical Oncology. 2010;**17**(11):2856-2862. DOI: 10.1245/s10434-010-1175-0. Epub 2010 Jun 22

[20] Stein SA, Bergamaschi R. Extracorporeal versus intra-corporeal ileocolic anastomosis. Techniques in Coloproctology. 2013;**17**(Suppl 1):S35-S39

[21] Hellan M, Anderson C, Pigazzi A. Extracorporeal versus intracorporeal anastomosis for laparoscopic right hemicolectomy. Journal of the Society of Laparoendoscopic Surgeons. 2009;**13**:312-317

[22] Jung KU, Park Y, Lee KY. Robotic transverse colectomy for mid-transverse colon cancer: Surgical techniques and oncologic outcomes. Journal of Robotic Surgery. 2015;**9**(2):131-136. DOI: 10.1007/s11701-015-0502-7. Epub 2015 Feb 17

[23] Reiley CE, Akinbiyi T, Burschka D, Chang DC, Okamura AM, Yuh DD. Effects of visual force feedback on robot-assisted surgical task performance. Journal of Thoracic and Cardiovascular Surgery. 2008;**135**(1):196-202. DOI: 10.1016/j.jtcvs.2007.08.043

Chapter 7

Robotic Hiatal Hernia Repair

Konstantinos Konstantinidis,
Michael Konstantinidis, Savvas Hirides and
Petros Hirides

Additional information is available at the end of the chapter

http://dx.doi.org/10.5772/intechopen.71164

Abstract

Robotic surgery has revolutionized medicine during the last 16 years by transformation of the classic operating theaters into computer-mediated working stations. Numerous procedures have been proved to be feasible and safe by using the continuously evolving, various robotic platforms. From the early beginnings of this revolution, challenging operations such as those concerning the gastroesophageal junction, especially in super-obese patients or during redo operations, proved out to have certain benefits when performed robotically, both for patients as well as for surgeons.

Keywords: robotic surgery, gastroesophageal reflux, cruroplasty, Nissen fundoplication, Toupet fundoplication

1. Introduction

From the early introduction of robotic surgical systems, upper gastrointestinal (GI) surgery has been one of the most promising areas of application. Numerous reports for successful robotic hiatal hernia and gastroesophageal reflux disease (GERD) surgery have been published [1–6]. Nissen fundoplication is the most commonly performed fundoplication. Partial fundoplications can be performed by adjusting the extent of the wrap. In any case, the main stages of the operation are performed according to the following description of the robotic Nissen fundoplication.

Indications:

- Symptomatic sliding hiatal hernia—GERD, esophagitis
- Paraesophageal hernia

Contraindications:

- Nonspecific
- Intolerance to anesthesia or laparoscopy. Bleeding tendency
- Relative: morbid obesity. Previous operations in the upper abdomen. Strictures from extensive esophagitis

Patient preparation:

- Gastrographin swallow
- Upper GI endoscopy (EGD)
- Esophageal manometry
- 24-h pH testing (not obligatory if patient presents with typical symptomatology)
- NPO for at least 8 h before the operation
- Admission at the day of surgery
- CXR, ECG, CBC, APTT, and INR at the day of surgery
- Preoperative antibiotic coverage (single dose at induction of anesthesia)

Operating room setup:

- da Vinci crew—technical support always necessary to be present
- Laparoscopic set availability (for the rare event of conversion)

Positioning of the patient and the robot:

- Anti-trendelenburg (**Image 1**)
- Robot comes in line with the camera port and the hiatus (**Image 2**)
- The surgeon should ensure continuous communication with the bedside assistants
- Bedside assistants should be experienced laparoscopic surgeons with certified training in the use of robotics

Pneumoperitoneum and trocar sites:

- 12 mm incision, 8 cm below the xiphoid and two fingerbreadths laterally to the midline (toward the left side of the patient)
- Pneumoperitoneum induction is done by using the Hasson technique. Alternatively, pneumoperitoneum may be induced by OptiView trocar (camera arm) by using 0° laparoscopic

http://dx.doi.org/10.5772/intechopen.71164

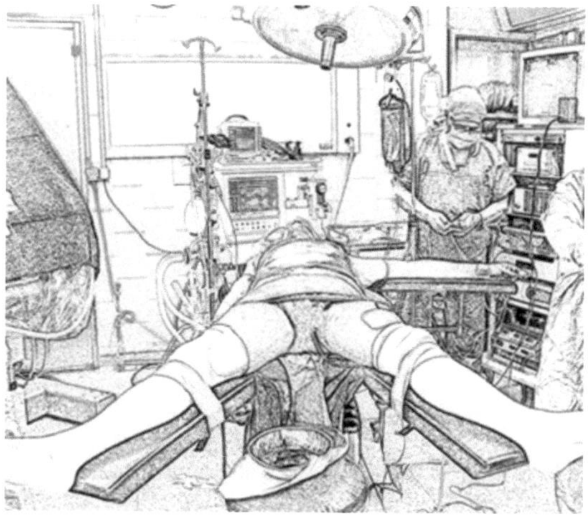

Image 1. Patient in supine position with legs apart.

Image 2. Positioning of the robot and team. The robot should be positioned at an axis created by the camera port and the hiatus. The surgeon should ensure continuous communication with the bedside assistants.

camera. In this case, make sure to recognize all layers of the anterior abdominal wall (subcutaneous fat, anterior sheath, muscular layer, and posterior sheath)

- Initial check of the abdomen to exclude other pathology can be performed with laparoscopic maneuvers, by holding the robotic camera and rotating to all four abdominal quadrants
- Three additional robotic trocars (8 mm) are inserted: left (Arm 1) and right (Arm 2) midclavicular and right anterior axillary line (fourth arm for retracting the liver). Incisions for trocars #1 and #2 should be done at least 3–4 cm below the costal margins and at an 8 cm distance from the camera port. A sterilized ruler may be used to confirm correct distance between ports

- One or even two (especially in the initial experience of the team) 5 mm assistant trocars can be added according to the needs of the operation. The first one is placed between camera port and trocar #1. The second one is place between camera port and trocar #2 (**Image 3a** and **3b**)

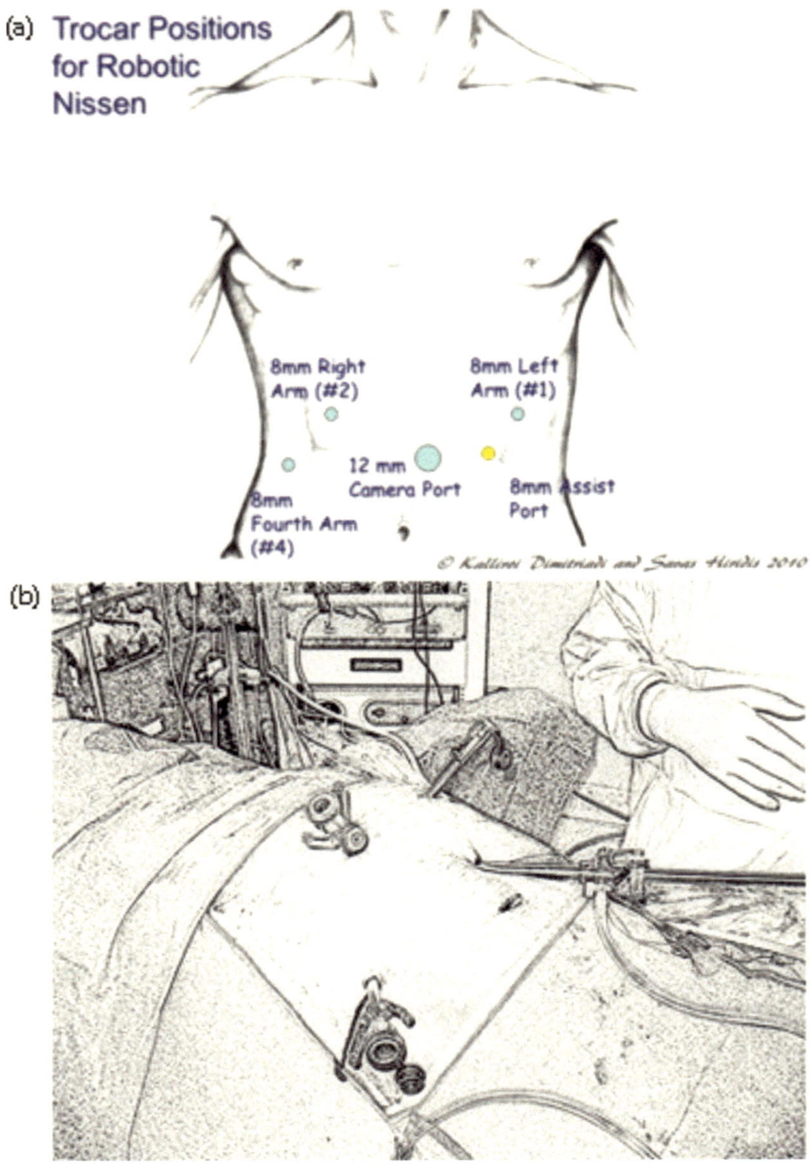

Image 3. (a) Trocar positions for robotic Nissen. (b) Initial check of the abdomen with conventional laparoscopy, using the robotic camera.

http://dx.doi.org/10.5772/intechopen.71164

2. Console setup parameters

In the present systems, setup of the console parameters remains quite simple and is usually done before the operation with the assistance of technical staff responsible for the system. The surgeon must adjust the position of his chair, his arm-rest and the lenses in order to achieve the optimal ergonomy. In the end, he can save these settings in his account, so that the system restores exactly the same position every time he logs in. Using the TilePro System, you may import images of patient's preoperative exams within the system for final considerations.

3. Stages of the procedure

3.1. Exposure

Install the liver retractor on Arm 4 and slowly retract the liver, exposing the gastroesophageal junction. Retraction of the liver is accomplished using Arm 4 with the Robotic Graptor. Alternatively, a bowel grasper can be used. Ask the bedside assistant to insert a laparoscopic grasper through the left lateral 5 mm port and retract the stomach laterally and inferiorly. This traction is mandatory throughout the whole procedure for the proper exposure of the gastroesophageal junction (**Image 4**).

3.2. Dissection in the lesser omentum

Install the Cadiere forceps on Arm 2 and the Monopolar Hook Cautery on Arm 1 (remember to use a reducer if the 5 mm Hook Cautery will be used). By gentle traction of the lesser omentum, create a window between the stomach and the liver edge (hepatogastric ligament), just

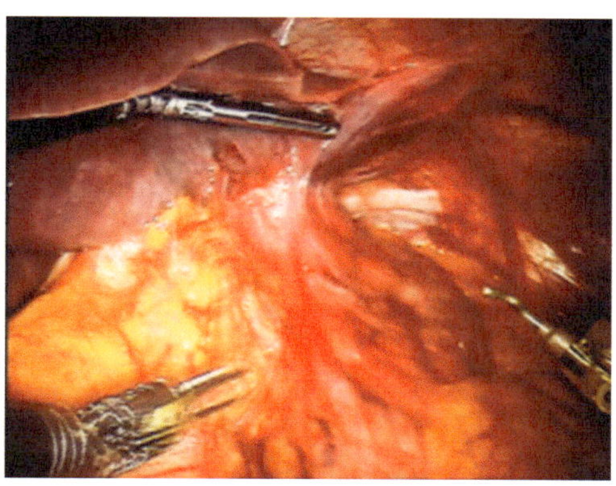

Image 4. Retraction of the liver using Arm 4 with the robotic Graptor. Retraction of the stomach is accomplished by a laparoscopic grasper from the bedside assistant.

above the caudate lobe of the liver. Beware to protect the right (hepatic) branch of the vagus nerve or any ectopic left hepatic arteries (that can be found next to the right branch), as you proceed proximally (**Image 5**).

3.3. Dissection at the crura and around the esophagus

As soon as the crural region was reached, careful dissection and stripping of the crura should take place. We usually dissect the right crus first. Dissection proceeds slowly with division of the superior portion of the phrenoesophageal ligament and toward the anterior surface of the left crus. Beware to protect the anterior branch of the vagus nerve, although stable traction ensured by the assistant trocar usually make it easily visible. Avoid grasping the esophagus at all times during the operation. Instead remember to ask for more traction on the stomach as the mobilization proceeds, exerted from the laparoscopic grasper of the bedside assistant. After complete dissection around the crura, mobilization of the esophagus is initiated by division of the numerous short adhesions to the crura. Extending this dissection as proximally as possible to ensure an adequate part of movable esophagus (at least 4 cm of esophagus should be able to move below the diaphragm without any tension). At this phase, a paraesophageal lipoma may be met, usually situated between the esophagus and the left crus. This is often rather voluminous and bleeds easily. Gently grasp with the robotic forceps (Arm #2) and pull back inside the abdominal cavity, while cauterizing any adhesions with the monopolar hook (Arm #1). After completion of the dissection, excise the lipoma and leave it under the liver but remember to remove before ending the operation (**Image 6**).

3.4. Creating the posterior window

Ask the bedside assistant to expose the angle between the right crus and the esophagus and start dissecting around the esophagus in a posterior direction. Take your time here because

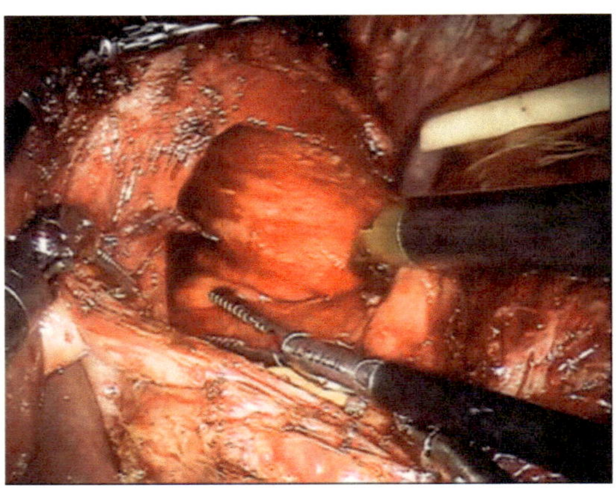

Image 5. View of the field after complete dissection of the hepatogastric ligament. On the left, the right crus is fully exposed. Inferiorly, the hepatic branch of the vagus nerve has been preserved.

apart from hurting esophagus itself, it is crucial to recognize and dissect the posterior branch of the vagus nerve at this point. Control any minor bleeding by using the robotic bipolar forceps. Avoid using monopolar for hemostasis (**Image 7**).

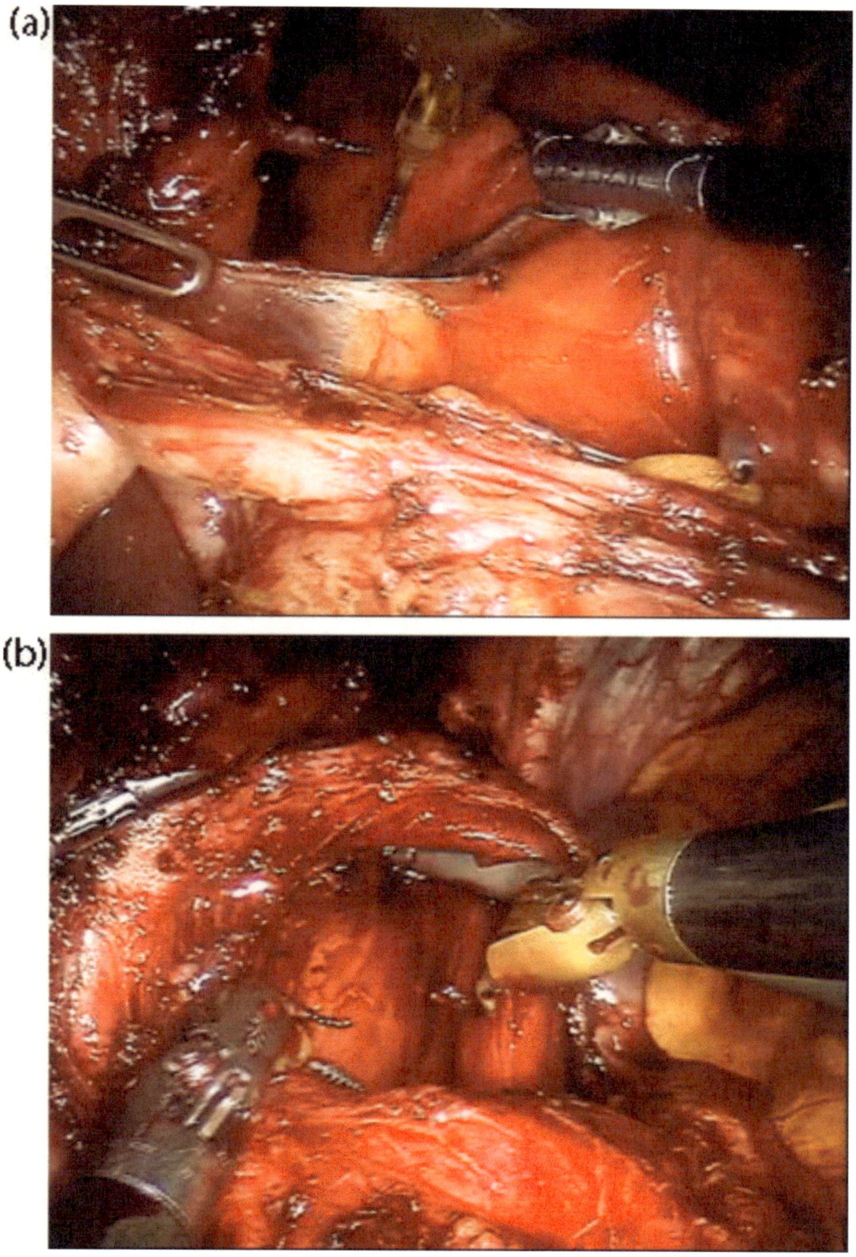

Image 6. (a) After complete dissection around the crura, mobilization of the esophagus is initiated by division of the numerous short adhesions to the crura. (b) Exposure of the left crus and dissection of its attachments to the lower esophagus.

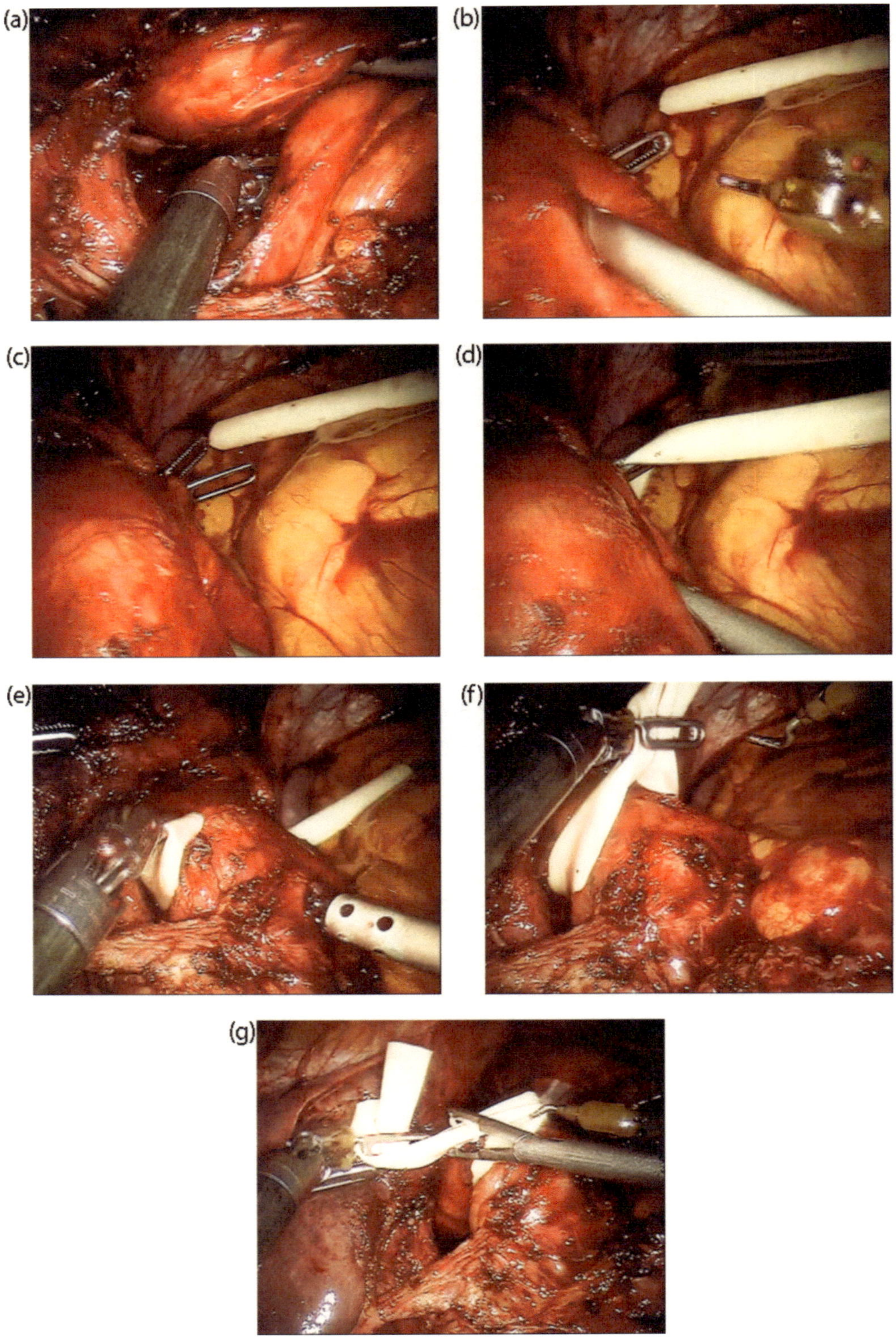

Image 7. (a-g): Snapshots from various phases of encircling the esophagus with a penrose drain by using the robotic grasper.

http://dx.doi.org/10.5772/intechopen.71164

At this point, ask the bedside assistant to introduce a short penrose drain through one of the robotic ports (usually #1). He should be aware that by this maneuver, pneumoperitoneum may be lost, so he must be fast but safe. Alternatively, he may use the 5 mm valve to introduce the drain without air loss.

Pass the robotic forceps slowly around the esophagus and grasp the penrose. By a backward movement, this should encircle the esophagus. The assistant secures the penrose with a hemolock clip and makes traction again by holding the penrose. Revise the crural dissection once again.

3.5. Division of short gastric veins

For adequate mobilization of the fundus, this is usually necessary. Your assistants should change the robotic monopolar with the robotic ultrasonic scissors (or Vessel Sealer) at this point. Approximately at one-third of the greater curvature length, make a window entering the omental bursa. Proceed cephalad with slow division of the short gastrics until the penrose drain at the gastroesophageal junction is met. Soft adhesions of the posterior gastric wall to the pancreas should be divided as needed. This part of the operation may be particularly troublesome and needs additional care as one proceeds proximally in tight proximity to the spleen, which can be easily injured. Use your second arm to gently retract the stomach and ask the bedside assistant to retract the omentum laterally. In this way, you should always find the correct plane to divide the short gastrics. In case of a minor hemorrhage, do not hesitate to put a sponge inside. This may immediately clean the field and help you identify the bleeding source (**Image 8**).

3.6. Suturing the crura

Now, proceed to close the defect of the hiatal hernia.

Many authors suggest that aNr.48-50 bougie should be in place while closing the crural defect. The authors have stopped using a bougie for Nissen fundoplications, early in their experience (**Image 9**).

Ask the assistant to pull the stomach laterally and superiorly in order to expose lower junction of the crura. Also, ask him to introduce a short piece of suture through port #1 (No. 2-0, nonabsorbable suture, 15 cm for every two stitches) and to change your robotic instruments with robotic needle holders. Suture the crura with thick bites and make sure to include the peritoneum to strengthen the suture. In addition, the use of pledgets is also advisable, especially in large defects. Use a figure-of-eight type of suturing. Usually 2–3 sutures are adequate. Robotic suturing is performed in an open-surgery fashion, that is, you simply rotate one arm around the other holding the end of the stitch. Laparoscopic suturing skills are not necessary for this phase of the operation.

3.7. Creating the wrap

Push part of the fundus toward the posterior window and then use the robotic forceps (Arm #2) to pull the fundus behind the esophagus. Do not try to do this in one step because the instrument may easily injure the gastric wall. Ask the bedside assistant to hold the fundus at this position and reposition your forceps by a larger (more secure) bite. Now pull the rest

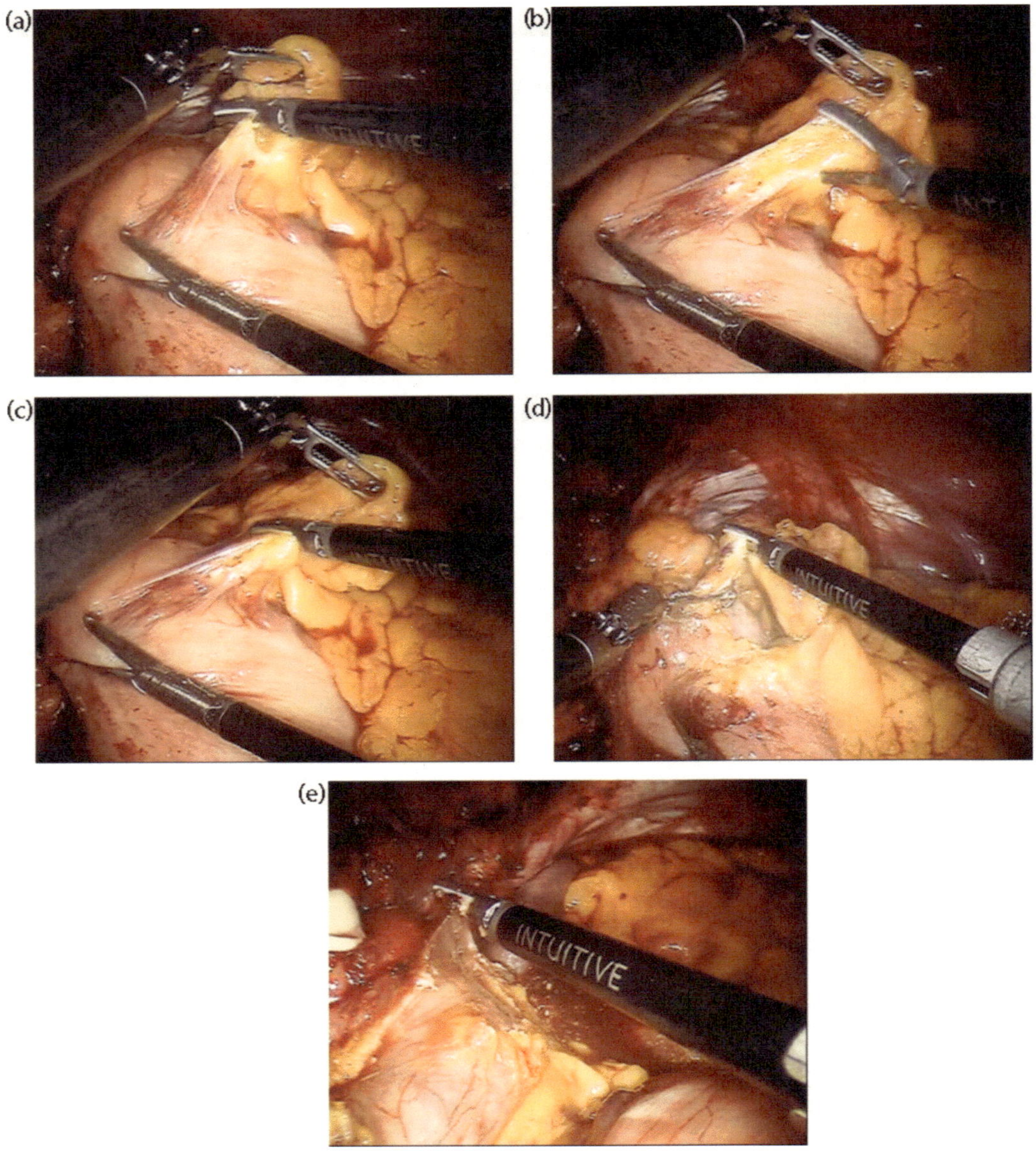

Image 8. (a-e): Snapshots from various phases of mobilizing the gastric fundus by division of the short gastrics. Note the difficulty as the dissection proceeds proximally and closer to the spleen.

of the fundus and bring it in front of the esophagus. You may assess tension of the wrap by gently pulling and pushing the fundus around the esophagus as you hold it at this point (shoe-shine maneuver). If your mobilization is adequate, the wrap should stay around the esophagus, else it may return at its initial position, outside the posterior window, which denotes that further posterior dissection may be necessary (**Image 10**).

http://dx.doi.org/10.5772/intechopen.71164

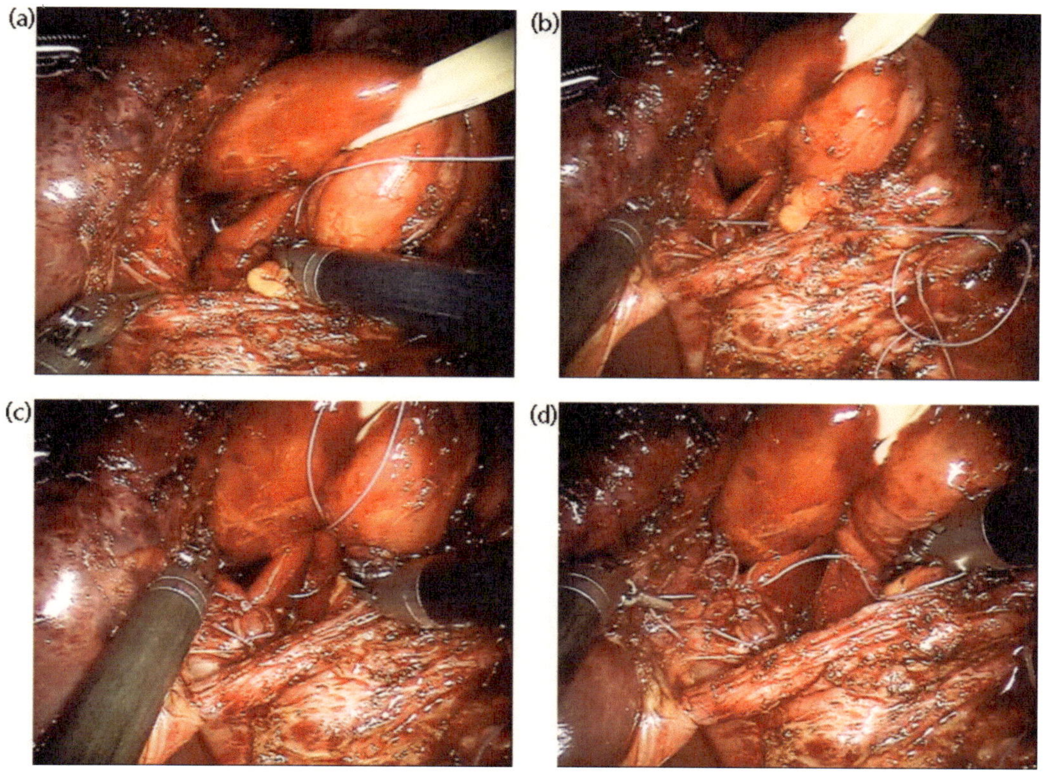

Image 9. (a-d): Snapshots from various phases of closing the defect by suturing the crura.

3.8. Anchoring the wrap

Assess the anterior surface of the stomach in order to anchor your wrap properly. Remove any large lipomas near the point of anchoring by using the ultrasonic scissors (or the newer Vessel Sealer). After that using the maneuvers described above, introduce once again a suture of the same nonabsorbable material and ask for the robotic needle holders in your hands. Approximate the left to the right part of the fundus and suture them together making a figure-of-eight stitch. These stitches should pass through all gastric wall layers and part of the anterior esophagus should be included with partial thickness bites. Many authors suggest securing the wrap to the diaphragm using two coronal sutures (left and right). This is not included in the standard technique of the authors. After completion of the anchoring, the assistant's 5 mm grasper should be able to pass below the wrap (maneuver to make sure that a "floppy Nissen" has been accomplished) (**Image 11**).

3.9. Final check and removal of instruments

Irrigation and suction is not necessary if no bleeding occurred during surgery. Remember to remove any material used during the procedure (failed clips, sponges, periesophageal

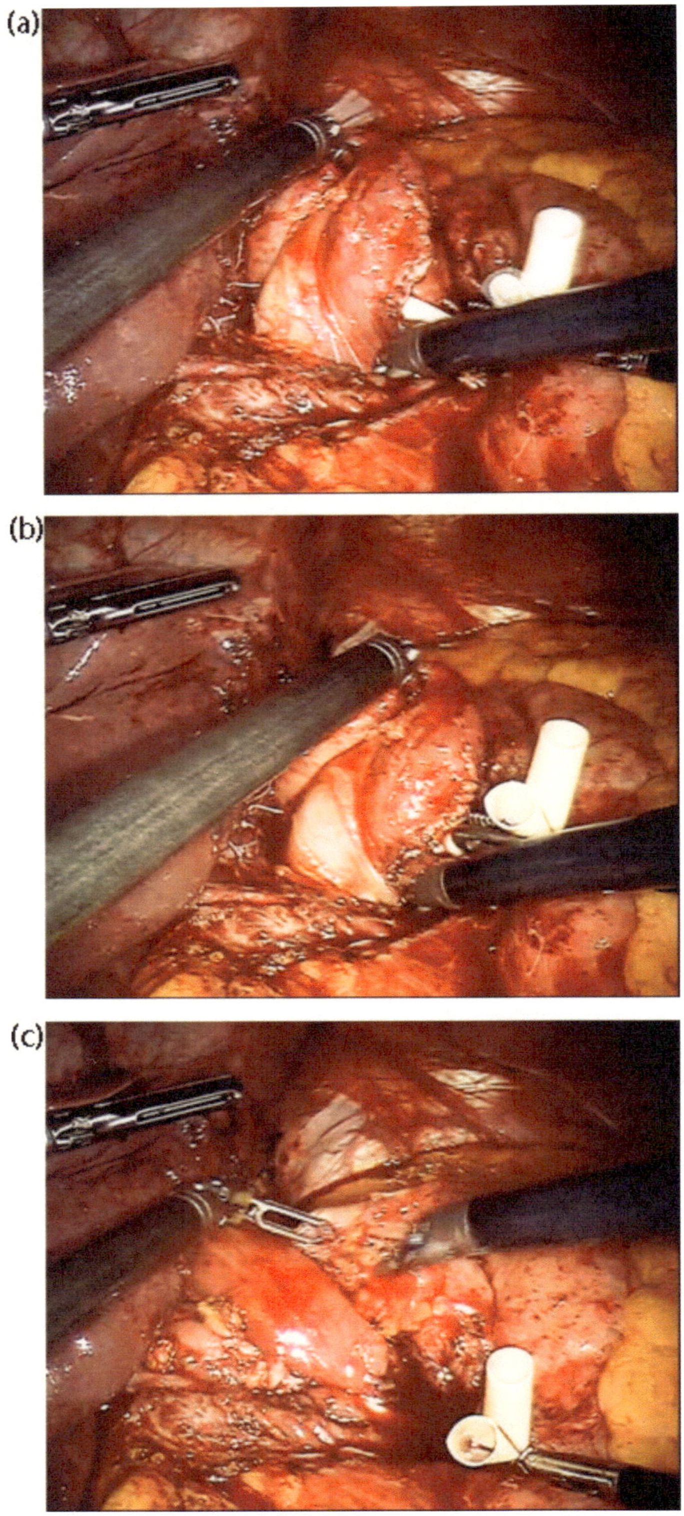

Image 10. (a-c): Mobilization is adequate because the wrap remains around the esophagus after cessation of the traction.

http://dx.doi.org/10.5772/intechopen.71164

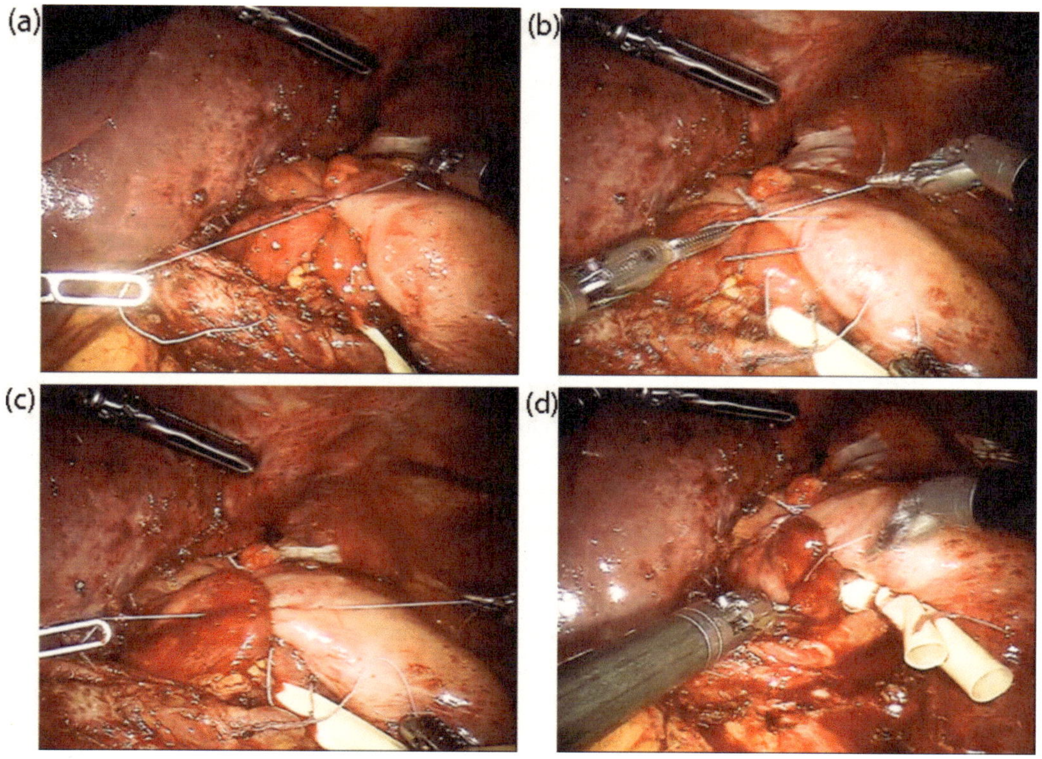

Image 11. (a-d): Snapshots from suturing the wrap for fixation in its final position.

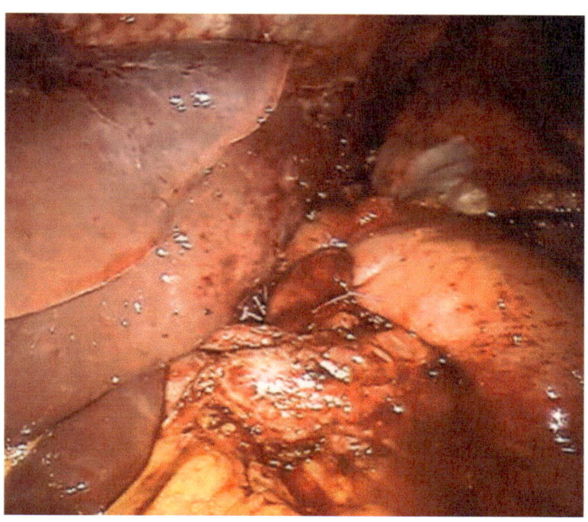

Image 12. Final result of the operation.

lipomas, lymph nodes, etc.) at this stage. Remove all instruments under direct vision, starting by the liver retractor which must be followed to the deep-seated fourth port in the right lower abdominal wall (**Image 12**).

3.10. Skin closure and wound dressings

If an OptiView technique was used for pneumoperitoneum at the beginning, there is no need for fascial closure. In case of open (Hasson) technique, a single figure-of-eight fascial suture is enough. Monofilament suture materials have been used to close the skin intradermally. Apply steri-strips and cover with water-resistant dressings.

4. Special considerations and hazards

- The large paraesophageal hernia

In the challenging case of large paraesophageal hernia, the technical difficulty of the operation rises significantly, and an experienced team should be called in. Soon after initial dissection at the crural region or even before this, assessment of the herniated content should be established. An effort to reduce the herniated viscera should be tried after complete adhesiolysis around the esophagus. Careful separation of the hernia sac and mobilization of the large accompanying lipoma should be anticipated in addition to the standard phases of the operation. Injury to the esophagus, to the vagal branches, or significant hemorrhage can occur during these stages.

- The short esophagus

Patients with advanced gastroesophageal reflux disease may present with a short esophagus. In practice, the surgeon should be able to differentiate between a truly short esophagus and an apparently short esophagus, which is more common and means that esophageal mobilization should be performed. The robotic system permits fine dissection in the narrow paraesophageal spaces even high in the mediastinum. Thus, the myth of a short esophagus should be treated with extensive mediastinal dissection of the lower esophagus (which according to the authors is usually enough), before a lengthening procedure is considered (Collis gastroplasty).

5. Postoperative management

- After completion of the operation, all port sites are injected with a solution of 20 ml of ropivacaine hydrochloride (2 mg/ml).
- The nasogastric tube is usually removed at the end of the operation.
- Normally, on the night of surgery, patients can receive oral fluids and should be mobilized.
- After a normal postoperative course, patients can usually be discharged within 48 h.
- Soft diet is suggested for the first 10 days after the operation.

6. Future perspectives—single-site robotic Nissen

Single-site robotic Nissen was reported using da Vinci straight instruments through laparoscopic single-site trocars [7, 8]. In May 2011, Konstantinidis et al. reported the first single-site robotic Nissen using the single-site curved instruments. The port was placed two fingerbreadths above the umbilicus and laterally to the midline. A cholecystectomy took place using the same trocar, before attempting the fundoplication. Although some exposure problems were recorded, the procedure was completed uneventfully. Single-incision surgery may prove to give an additional benefit to the use of robotics in surgery by providing steady three-dimensional image and intuitive instrumentation through a single 2.5 cm incision. Results of robotic fundoplication have been promising from numerous studies [9–15]; but up to now, supporters failed to publish an evidence-based proof of its superiority versus existing laparoscopic techniques [16–18].

Author details

Konstantinos Konstantinidis, Michael Konstantinidis, Savvas Hirides* and Petros Hirides

*Address all correspondence to: savashiridis@yahoo.com

Athens Medical Center, Greece

References

[1] Margaron FC, Oiticica C, Lanning DA. Robotic-assisted laparoscopic Nissen fundoplication with gastrostomy preservation in neurologically impaired children. Journal of Laparoendoscopic & Advanced Surgical Techniques Part A. 2010 Jun;**20**(5):489-492. PMID:20518687[PubMed - indexed for MEDLINE]

[2] DeUgarte DA, Hirschl RB, Geiger JD. Robotic repair of congenital paraesophageal hiatal hernia. Journal of Laparoendoscopic & Advanced Surgical Techniques Part A. 2009 Apr;**19**(Suppl 1):S187-S189. PMID:19331624[PubMed - indexed for MEDLINE]

[3] Draaisma WA, Gooszen HG, Consten EC, Broeders IA. Mid-term results of robot-assisted laparoscopic repair of large hiatal hernia: a symptomatic and radiological prospective cohort study. Surgical Technology International. 2008;**17**:165-170. PMID:18802897[PubMed - indexed for MEDLINE]

[4] Hartmann J, Jacobi CA, Menenakos C, Ismail M, Braumann C. Surgical treatment of gastroesophageal reflux disease and upside-down stomach using the Da Vinci robotic system. A prospective study. Journal of Gastrointestinal Surgery. 2008 Mar;**12**(3):504-509. Epub 2007 Nov 20.PMID:18027060[PubMed - indexed for MEDLINE]

[5] Braumann C, Menenakos C, Rueckert JC, Mueller JM, Jacobi CA. Computer-assisted laparoscopic repair of "upside-down" stomach with the Da Vinci system. Surgical Laparoscopy Endoscopy & Percutaneous Techniques. 2005 Sep;**15**(5):285-289. PMID:16215489[PubMed - indexed for MEDLINE]

[6] Hanisch E, Markus B, Gutt C, Schmandra TC, Encke A. [Robot-assisted laparoscopic cholecystectomy and fundoplication—Initial experiences with the Da Vinci system]. Chirurg. 2001 Mar;**72**(3):286-288. German. PMID:11317449[PubMed - indexed for MEDLINE]

[7] Markar SR, Karthikesalingam AP, Hagen ME, Talamini M, Horgan S, Wagner OJ. Robotic vs. laparoscopic Nissen fundoplication for gastro-oesophageal reflux disease: systematic review and meta-analysis. The International Journal of Medical Robotics. 2010 Jun;**6**(2):125-131. Review. PMID:20506440[PubMed - indexed for MEDLINE]

[8] Allemann P, Leroy J, Asakuma M, Al Abeidi F, Dallemagne B, Marescaux J. Robotics may overcome technical limitations of single-trocar surgery: An experimental prospective study of Nissen fundoplication. Archives of Surgery. 2010 Mar;**145**(3):267-271. PMID:20231627[PubMed - indexed for MEDLINE] Free Article

[9] Ceccarelli G, Patriti A, Biancafarina A, Spaziani A, Bartoli A, Bellochi R, CasciolaL. Intraoperative and postoperative outcome of robot-assisted and traditional laparoscopic Nissen fundoplication 1. European Surgical Research. 2009;**43**(2):198-203. Epub 2009 Jun 10

[10] Albassam AA, Mallick MS, Gado A, Shoukry M. Nissen fundoplication, robotic-assisted versus laparoscopic procedure: A comparative study in children. European Journal of Pediatric Surgery. 2009 Oct;**19**(5):316-319. Epub 2009 Jun 9. PMID:19513968[PubMed - indexed for MEDLINE]

[11] Müller-Stich BP, Reiter MA, Mehrabi A, Wente MN, Fischer L, Köninger J, Gutt CN. No relevant difference in quality of life and functional outcome at 12 months' follow-up-a randomised controlled trial comparing robot-assisted versus conventional laparoscopic Nissen fundoplication. Langenbeck's Archives of Surgery. 2009 May;**394**(3):441-446. Epub 2009 Jan 23. PMID:19165497[PubMed - indexed for MEDLINE]

[12] Copeland DR, Boneti C, Kokoska ER, Jackson RJ, Smith SD. Evaluation of initial experience and comparison of the da Vinci surgical system with established laparoscopic and open pediatric Nissen fundoplication surgery. Journal of the Society of Laparoendoscopic Surgeons. 2008 Jul-Sep;**12**(3):238-240.PMID:18765044[PubMed - indexed for MEDLINE] Free PMC Article

[13] Meehan JJ, Meehan TD, Sandler A. Robotic fundoplication in children: resident teaching and a single institutional review of our first 50 patients. J Pediatr Surg. 2007 Dec;**42**(12): 2022-2025. PMID:18082700[PubMed - indexed for MEDLINE]

[14] Müller-Stich BP, Reiter MA, Wente MN, Bintintan VV, Köninger J, Büchler MW, Gutt CN. Robot-assisted versus conventional laparoscopic fundoplication: Short-term outcome of a pilot randomized controlled trial. Surgical Endoscopy. 2007 Oct;**21**(10):1800-1805. Epub 2007 Mar 13. PMID:17353978[PubMed - indexed for MEDLINE]

[15] Heemskerk J, van Gemert WG, Greve JW, Bouvy ND. Robot-assisted versus conventional laparoscopic Nissen fundoplication: A comparative retrospective study on costs and time consumption. Surgical Laparoscopy Endoscopy & Percutaneous Techniques. 2007 Feb;**17**(1):1-4. PMID:17318044[PubMed - indexed for MEDLINE]

[16] Draaisma WA, Ruurda JP, Scheffer RC, Simmermacher RK, Gooszen HG, Rijnhart-de Jong HG, Buskens E, Broeders IA. Randomized clinical trial of standard laparoscopic versus robot-assisted laparoscopic Nissen fundoplication for gastro-oesophageal reflux disease. The British Journal of Surgery. 2006 Nov;**93**(11):1351-1359

[17] Nakadi IE, Mélot C, Closset J, DeMoor V, Bétroune K, Feron P, Lingier P, Gelin M. Evaluation of da Vinci Nissen fundoplication clinical results and cost minimization. World Journal of Surgery. 2006 Jun;**30**(6):1050-1054

[18] Morino M, Pellegrino L, Giaccone C, Garrone C, Rebecchi F. Randomized clinical trial of robot-assisted versus laparoscopic Nissen fundoplication. The British Journal of Surgery. 2006 May;**93**(5):553-558